THE

Miner's Canary

THE

Miner's Canary

UNRAVELING THE
MYSTERIES OF EXTINCTION

NILES ELDREDGE

Princeton University Press
Princeton, New Jersey

Published by Princeton University Press, 41 William Street,
Princeton, New Jersey 08540

Copyright © 1991 by Niles Eldredge

Library of Congress Cataloging-in-Publication Data

Eldredge, Niles.
    The miner's canary: unraveling the mysteries of extinction /
Niles Eldredge
       p.   cm.—(Princeton science library)
    Originally published: New York : Prentice Hall Press, © 1991
    Includes bibliographical references and index.
    ISBN 0-691-03655-1 (pbk. : alk. paper)
    1. Extinction (Biology) 2. Ecology. 3. Biological diversity.
I. Title. II. Series.
[QE721.2.E97E57 1994]
575'.7—dc20   93-50185

First Princeton Science Library printing, 1994
Reprinted by arrangement with Simon & Schuster, Inc.

Princeton University Press books are printed on acid-free paper
and meet the guidelines for permanence and durability of the
Committee on Production Guidelines for Book Longevity
of the Council on Library Resources

10 9 8 7 6 5 4 3 2 1

Printed in the United States of America

For Norman D. Newell, who taught us the reality of mass extinction in the geological past. And for all who seek to mitigate the impact of *Homo sapiens* on the ecosystems of our planet.

Acknowledgments

I am grateful to my colleagues at the American Museum of Natural History, many of whom supplied valuable information, and all of whom provided much inspiration through their dedication to the living world, both ancient and modern. I especially thank Ian Tattersall for initiating me into the Madagascar experience, both in the field and back in New York—and for his many ecological and evolutionary insights.

Contents

ix

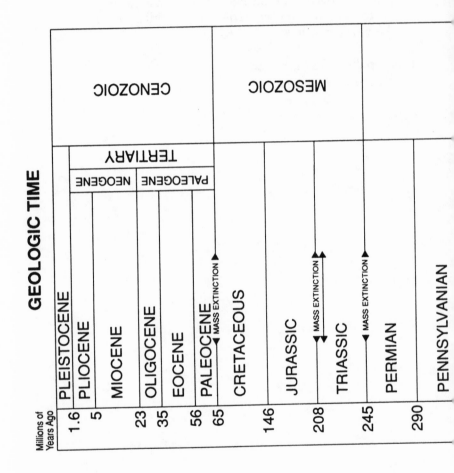

GEOLOGIC TIME

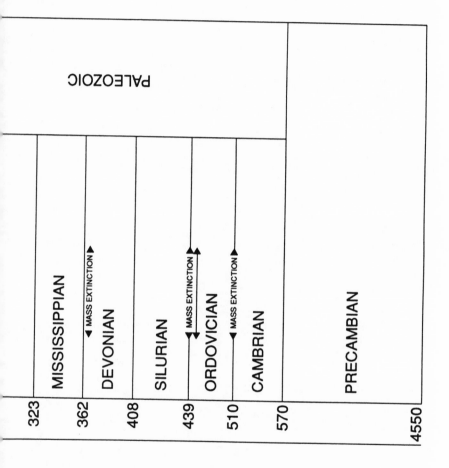

Prologue

In the days before sophisticated sensing technology, miners often took canaries with them as they ventured down into the bowels of the earth. The canaries were their early warning system for poisonous gases. Simple and expedient, if a bit grisly, the health of the canaries mirrored the health hazards posed to the miners themselves.

Humans have long been concerned with the state of their ambient surroundings. In our transition from members of local ecosystems to our newly developed status as a far-flung species interacting in a very real sense with the earth's total ecosystem, we continue to worry about the state of our local environment. But we also see the effects, too often negative, that we seem collectively to be having on fundamental features of the global natural economy. We destroy habitats for our own short-term economic good. We change the proportion of gases in the atmosphere, decreasing ozone and raising global temperatures. We even single out particular prey items for overexploitation, driving them to—and sometimes over—the brink of extinction.

All we have to do is check the skies every spring to see the ~Rachel~ drastic decline of songbirds who leave their ever-diminishing winter homes in tropical forests to brave their ever-more hazardous migration routes up to their increasingly sparse and degraded breeding grounds in the northern hemisphere. Migrating songbirds fill the same role on a global scale that those caged canaries used to perform for miners. By now, it is abundantly clear to all who will look that the global miner's canary is not at all well.

The alarm over the state of the environment is nothing new, of course. Each generation has produced its share of observers echoing the theme of "things ain't what they used to be." Some simply detest change of any kind. We need not listen to them. But others have pointed to changes which, by any standard, must be seen as negative—ultimately, as threats to our own collective continued existence. In the crescendo of recent concern over the state of the global environment, I have encountered no words more trenchant and telling than the elegantly crafted arguments for conservation penned by Aldo Leopold in his 1949 classic *A Sand County Almanac*. All the reasons for alarm, and all the reasons (ranging from the esthetic to the utilitarian) for trying to salvage what is left of the earth's richness in the diversity of life, that we commonly hear these days can be found in Leopold's slim volume. The only difference is that far more people have become alarmed. The Green Party is an active part of the political process of many countries now—perhaps the most significant sign of heightened awareness of the state of, and a willingness to do something about, our environment.

I am a paleontologist by training and past experience. I am used to immersing myself in the details of the anatomy, evolution, classification, and ecology of creatures that lived 380 million years ago. All of the species I have studied are, of course, long gone. Indeed, the large group to which I have devoted most of my research time—trilobites, primitive arthropod kin to shrimps and insects—is itself totally extinct. Trilobites have been extinct, as far as anyone knows, for the last 240 million years!

Extinction to a paleontologist is a fact of nature. For years, I paid no particular attention to extinction, greatly preferring to emphasize the positive side of my data: what trilobites can tell me about the nature of the evolutionary process. But I am also a living, breathing citizen of this planet. I cannot help but sense the growing alarm that my colleagues at the American Museum of Natural History have over the state of the world's local ecosystems—leading to the loss of species after species of their favorite creatures. And, as an ordinary citizen, I cannot help but react to the nonstop media bombardment, the almost constant stream of unrelievedly bad news about the state of the very local environment in which I live with my family and friends.

I have taken to spending as much time these days tramping fields, forests, and wetlands looking at birds as I do hunting and studying trilobites. I look at birds for a complex set of reasons, with pure enjoyment and relaxation heading the list. But the world's birds, especially those that are, or *ought to be*, in my backyard, have quickly become my personal miner's canary.

I have come to realize that these two separate threads—the remote past and the modern world—really have much to reveal to one another. They are simply different strands of the same rope. Extinction—truly massive, global extinction—is indeed a fact of the history of life. Thanks especially to the demise of the dinosaurs some 65 million years ago, and thanks, too, to all the attention to the bold hypothesis claiming that it was a disastrous collision between the earth and an extraterrestrial asteroid or comet that did the dinosaurs in, the public at large is at least passingly familiar with the idea of mass extinction. We are also all more or less aware that species are disappearing at an alarming rate right now: Extinction is a fact of life in the modern world.

But what, really, causes extinction? Because extinction is the loss of biological diversity, that is, numbers of species, we need to ask: What generates and maintains patterns of diversity in the first place? Why are there so many species of beetles, and not so many species of primates? Why are there more species represented on an acre of land in the tropics than on the tundra? Do the natural processes that build and maintain these diversity patterns have anything to do with their collapse in spasms of extinction? And what about our readiness to shoulder virtually all the blame for the current state of environmental affairs? After all, mass extinctions have occurred sporadically over the last 3.5 billion years, and creatures that can be legitimately called "human" have only been around for 150,000 years (the age of our species, *Homo sapiens*), or, stretching things a bit to include extinct species of our own lineage, perhaps as much as 4 or 5 million years. Can we pinpoint what causes extinction without human intervention, then see what happens when we add humans to the mix? Can we come up with a general theory of extinction that relates past to present, and perhaps helps us see a bit more clearly the nature of our own present-day situation?

I think we can. I set out to write this book with the goal (or, really, simply the hope) of finding a simple, overarching theory of extinction. I am convinced that I have found it. It is woven from many familiar threads. It sees ecosystem collapse and the extinction of species as the outcome, generally, of earthbound causes. These causes are simply the flip side of the same factors that build and maintain ecosystems and lead to the evolution of new species in the first place. And there is indeed a role that humans (as we already know, anyway) are playing in this scenario: Our activities perturb the system along ways that are common to the great majority of mass extinction events that have occurred without our help in the geological past.

Most of us do not carry a chart of geological time around in our heads. The geological time scale, as we shall see, is based to a great degree on the occurrence of mass extinctions in the remote past. You will find a chart of geological time on pages xii-xiii, to which you may refer as needed in the course of our survey of mass extinctions.

I invite you to come along as we try to unravel the mysteries of extinction—and, in the process, to explore the ways in which the natural world is put together, and how life has managed to make it to the twentieth century. The exploration sheds a goodly measure of light on what lies in store for the twenty-first century.

The Miner's Canary

CHAPTER ONE

Extinctions Are for Real

FORTY MILES or so south of Copenhagen lies the sleepy hamlet of Stevn's Klint. A quaint little museum stuffed with an eclectic assortment of odds and ends from the past few centuries stands out among the open fields and the few simple houses. That is all, save the ruins of an old church now perched precariously close to the edge of the cliff overlooking the Baltic. The church is built from rock taken from those eroding cliffs—chalky limestone itself composed of the debris of the ages.

The chalk is relatively young as these things go, around 65 million years old. The sturdy wooden staircase leading down to the rocky shore below takes you back through time, though no one knows exactly how much time is represented in the 200 feet of limestone exposed in the cliff face. Chalk deposits accumulate slowly. Their major elements are the ultramicroscopic calcareous plates of single-celled, armored algae—minute photosynthesizing organisms floating in the plankton near the water's surface when living, sinking to the bottom when dead. How much time in 200 feet of chalk and chalky limestone? Over a million years, certainly. And probably much more—as much, say, as 3 or 4 million years.

But however much time may be there in the cliffs at Stevn's Klint, one thing is certain: During the interval when those sediments were quietly accumulating, something major happened. Something that figuratively—and some say literally—shook the earth. For midway down that cliff face, recorded in a thin, discontinuous band of green-gray clay no more than an

I

inch or so at its thickest, lies that point in time when the complexion of life took a dramatic change. Not just life in what is now northeastern Denmark but life all over the earth. Not just life under oceanic waves but life on land as well. That little layer of clay (the Danes call it "fish clay" because it has a few bones and scales in it) marks one of the greatest transitions in the entire history of life.

A shocking number of species alive before that clay was formed disappeared forever, their places taken by other species in the new ecosystems that came along after the fish clay. The fish clay holds within it the chemical clues that suggest that the earth very likely collided with an extraterrestrial body, either a meteor or a comet. Though we cannot see the dramatic effects of extinction in the fossil-poor rocks below that little layer of green clay, it is the clay itself that has helped focus so much attention on mass extinctions of the past and has sparked so much debate on the causes of what amount to worldwide nearly total ecosystem collapse.

Prior to this big extinction 65 million years ago, the earth had already seen a number of critical moments in life's history, most notably life's very invention at least 3.5 billion years ago. Complex cells came along 2 billion years later. Then there was the sudden rise of complex forms of life nearly 600 million years ago. After that we come to the greatest of all revolutions yet in the course of life's history: the grand extinction 245 million years ago at the end of the Permian Period that saw a dramatic close to the Paleozoic and consequent beginning to the Mesozoic Era. Much later, here at Stevn's Klint, lie sediments deposited at the very end of the Cretaceous Period—at the end of the Mesozoic and at the very beginnings of the Cenozoic, the time of modern life, the period of geological time when mammals became dominant, eventually *our* time. Before that, through the 180 million years of the Mesozoic, dinosaurs and various scaly congeners ruled the terrestrial roost, and life in the sea had a rather different complexion than it does today.

Fossils, at least of the large easy-to-spot-and-collect variety, like clams and sea urchins, are not all that common at Stevn's Klint, either above or below the notorious fish clay. Casual visitors, even if well versed in techniques of fossil collecting, would never guess that the cliffs preserve a momentous and critical event in the history of life. The limy sea bottoms that ultimately become chalk deposits are often poor places for animals and plants to live, and are sometimes inhospitable to the point of being watery equivalents of arid deserts. But geologists have known for more than 150 years that something happened that transformed the complexion of life on both land and sea: a transition that always seems, wherever we see its traces, pretty abrupt. It was a transition that forever changed the composition of terrestrial animal and plant communities, as well as those that, like Stevn's Klint, were developed beneath the waves. When we piece together all the evidence, we find that life was changed rather radically during the time that the fish clay layer at Stevn's Klint was forming.

We have known that extinctions have been both commonplace and critical shapers of the course of life's history for a very long time now. Baron Georges Cuvier, a nobleman who managed to keep both his head and professional life intact through the French Revolution, first published his treatise *Discours sur les Revolutions de la Surface du Globe* in 1812. The father of vertebrate paleontology, Cuvier was an anatomist who extended his studies to the fossil vertebrates that were commonly encountered in the quarries in the Parisian environs. With Alexandre Brongniart, Cuvier published one of the earliest of geological maps. Such maps depict the distribution of rocks of a region, and critical to any such mapping exercise is some notion of how rocks in different places match up, or are the same in one way or another. After all, most places are not like the Grand Canyon, with its broad stretches of exposed strata. Rather, it is one outcrop of sandstone on a hilltop here, one exposure of shale in a creek bed there, and an exposure of limestone beds in a cement quarry in still another spot.

One way to make a map is simply to connect all the spots that have similar rock types: all the sandstone, all the shale, all the limestone. The problem with that technique, though, is that sedimentary rocks come in layers, piled one on top of the other. Unlike the rather monotonous sequence of chalk at Stevn's Klint, the usual case is to find a variegated series of layers, such as shale on top of sandstone, perhaps with limestone on top of the shale, followed by more shale, with sandstone once again at the top of the cliff. The same rock type occurs in different layers, over and over again, as the sequence of layered rocks is traced from bottom to top in any particular region.

Although a number of the ancients appreciated the overall nature of bedded sedimentary rocks, it was really the Danish physician Niels Steensen (Nicolaus Steno in Latinized form) who formulated the simple principles that are the very key to deciphering earth history. Realizing that sandstones, limestones, and shales are simply the hardened versions of sand, lime, and clayey mud deposits, and knowing, too, that rivers are constantly transporting sediments, dumping them in lakes, river mouths, and seaways, Steno saw that layers at the bottom of a rock pile must have been deposited first. The upper layers must come later in time. Thus was the first explicit connection made between rock layers and notions of time in earth history. For ever after, it has been the geologist's goal to match up not just similar rock types but rock layers that are equivalent in time—layers exposed in different places that were formed at about the same time.

Cuvier and Brongniart mapped the rock strata (mostly limestones and chalks) around Paris according to their interpretation of the ages of the rocks. They had to have a way of telling time, of deciding which rock layers were equivalent to which. The technique they hit on was simplicity itself: They realized that the fossils that were turning up in these rocks in great numbers always occurred in the same general sequence. They realized that these fossils were of extinct organisms—both marine and terrestrial animals no longer known to inhabit the earth. They also realized, as had Linnaeus and other naturalists before them, that fossils could be classified just as readily as modern-day organisms. They needed but one further deductive inference: that fossils of the same species, found in different quarries throughout the Parisian region, must be the remains of animals alive at about the same time. Similar fossils, they reasoned, imply equivalent ages.

But that was not all. Cuvier's *Revolutions* were not just a projection of recent French political happenings into biological history. Cuvier's reading of Parisian biological history convinced him that life's history came packaged in a series of multiple creations, each ending in a catastrophic purge that set the scene for the next, subsequent creation. He thought he could specify more than thirty such revolutions in all.

The first half of the nineteenth century saw a virtual explosion of work, as geologists collected fossils, made field observations, correlated their rocks, and laid out the basic structure of our modern geological time scale. The earliest work, almost as a matter of course, was in Europe, following the pioneering efforts of Cuvier and Brongniart in France, and of William Smith in England. By midcentury, the major divisions of the geological time scale were essentially established.

We still use these names that bedevil beginning geology students as they grope for mnemonic devices (many of them earthily salacious) to master Cambrian, Ordovician, Silurian, Carboniferous, Permian, Triassic, Jurassic, Cretaceous, Paleocene, Eocene, Oligocene, Miocene, Pliocene, Pleistocene, Recent. We still use Paleozoic to embrace the first five of those names. Mesozoic takes in the next three—Triassic, Jurassic, Cretaceous—the vast middle age of life, the time of the dinosaurs, the era that ended midway up that chalk cliff at Stevn's Klint. Cenozoic covers from Paleocene to Recent (or Holocene, meaning "thoroughly Recent").

Nowadays, we have some hard numbers to tack onto the beginnings and endings of these various subdivisions of geological time. We know, for example, that the Cambrian began around 570 million years ago. The Paleozoic ended at the close of its last period, the Permian, about 245 million years ago—the approximate start of the Mesozoic. The Cretaceous ended 65 million years ago. We also know that the earth was formed somewhere on the order of 4.5 billion years ago. And life is at least as old as 3.5 billion years—a minimum estimate, as this is simply the age of the oldest fossils (bacteria) yet found.

But for the first 4 billion years of earth history there are relatively few subdivisions of time to plague novices. The reason there are so many names to learn on the geological time scale during only the last one-ninth of geological time is that it is precisely how long macroscopic life (multicellular plants and animals you can see without a microscope) has been a conspicuous presence, forming all those ecosystems in all the major physical environments of the earth's surface. All those divisions are really more appropriately labeled Cuvierian packages—divisions of geological time with beginnings and ends, marked by characteristic faunas and floras that typify the time; faunas and floras that were not separately created, as Cuvier had thought, but were derived from remnants of the previous package of living systems; packages of life that were destined to disappear in turn, passing on the flame from their own remnants to sow the seed for the next evolutionary and ecological diversification, the next peopling of the globe. For Cuvier was surely right, in the main: Life has indeed come in packages, biological systems that diversify and flourish over the earth's surface, only to vanish and be replaced by a succeeding system.

Extinctions are for real. They have had an exceedingly practical side effect: We can tell geological time because they have occurred. Creationists, who attribute life's history to the actions of a supernatural creator, are fond of accusing geologists of devising the geological time scale as a device to support the very idea of evolution. Nothing could be further from the truth: Cuvier and many of his early colleagues were creationists. Darwin was born three years before Cuvier's *Revolutions* was published; by the time *The Origin of Species* was published in 1859, nearly all the elements of our modern time scale were in place, thanks to the diligent efforts of these early creation-minded geologists who simply documented the sequence of Cuvierian life packages and thus revealed not only the general sequence of earth history but the history of life itself. And that history has been, as Cuvier first pointed out so long ago, shot through with revolutions—revolutions engendered by extinctions that periodically upset the status quo.

Some extinctions, as one might imagine, have been more devastating, more nearly all-encompassing, than others. Once again, we need only consult a chart of geological time to see that this is so. Take, for example, the three major divisions (eras) of Phanerozoic time: the Paleozoic, Mesozoic, and Cenozoic. Their names say it: *Phanerozoic* means visible life—when life ceased being restricted to single microscopic cells (bacteria, algae, and protozoans) and included bigger things such as jellyfish, various worms, vertebrates, mollusks, vascular plants, and arthropods. *Paleo* means ancient; *Meso*, middle; and *Ceno*, modern. It will perhaps come as no great surprise that two of the biggest extinction events (yet) in life's history form the dividing line between Paleozoic and Mesozoic (the Great Extinction at the end of the Permian, most devastating of all, when more than 90 percent, and maybe even 96 percent, of all species on earth suffered extinction) and between Mesozoic and Cenozoic (the great Cretaceous event recorded at Stevn's Klint and many other places around the globe).

These were globally encompassing events ending one era and beginning another. It turns out that the more nearly global, the more ecologically far-flung, and the more taxonomically encompassing (meaning the more that different kinds of life forms are affected) the extinction, the greater the differences between the ecosystems that disappear and those that eventually come to take their place. Extinction has a profound effect on the course of evolution. The greater the extinction, the more radically different the kinds of organisms tend to be that evolve to take the place of those that have disappeared.

Evolution depends very much on extinction; it is almost as if extinction plays a creative role in the evolutionary process.

Such a message can be more than a little disturbing. As we shall see, it has been used as an excuse *not* to intervene to try to stem the tide of modern extinctions. The argument goes something like this: Extinctions are natural phenomena, and thus far new species have always evolved to take the place of their fallen predecessors. We need not, or so it has been said, fear the loss of species around us as they, too, are destined to be replaced in the long run. (I will have a good deal more to say on this train of thought later.) In contrast, a number of leading conservationists have recently been advocating measures to mitigate our current problems with extinction on precisely the opposite claim: Extinction, they say, will inhibit further evolution, as it removes genetic information that is essential for future evolution to occur.

This issue is one of the trickiest we face when coming to grips with the nature—and significance—of extinction. It will come into better focus as we delve more deeply into the nature of extinction: what causes it, what the structure of biological systems may be that both fosters stability and change. But, for the moment, we need to balance the claim that extinction is absolutely vital for evolution to occur with the equally compelling claim that loss of genetic information through extinction severely limits future evolutionary possibilities.

Evolution (at least to those not opposed to the very idea on ideological grounds) often seems to be a good thing. Specifically, evolution, ever since its early days of acceptance in the mid-nineteenth century, has been closely associated with the idea of progress. "Change is inevitable in a progressive country," wrote Benjamin Disraeli in 1867. *Progress, change, improvement:* These Victorian buzzwords both reinforced and were reinforced by the Darwinian notion of evolution through natural selection. Darwin himself visualized evolution primarily as a matter of slow, steady, gradual change, even improvement, making the explicit analogy with the efforts of breeders to improve the various qualities of their domestic pigeons, dogs, cattle, and agricultural plants. What we can achieve through selective breeding over but a few generations is nothing compared to what nature can achieve through competition for limited resources over millions of years, or so went the Darwinian argument.

Nor is there any serious doubt that natural selection is a potent force in nature. Organisms do indeed vary from one to another: Some are better than others at coping, extracting energy from the environment, growing, and maintaining themselves. On the whole, these more vigorous individuals, those who lead more successful economic lives, will also tend to leave more offspring behind. Because organisms tend to resemble their parents (we now have a much clearer idea than Darwin had about the principles of genetics), those traits that gave the parents an advantage in the economic game of life will tend to be passed along to their offspring, just as Darwin saw must happen as a matter of course.

But, more recently, we are coming to realize that the powerful mechanism of natural selection does not by any means imply that drastic, if gradational, change is inevitable as the geological ages role by. Organisms are masters at finding suitable habitat. When the environment changes in one place, organisms will chase it. If they are fish, they will swim to it. If they are plants, their seeds will find it. That's how most species of American forest trees survived as the glaciers came down through what is now the northern United States no fewer than four times during the last 1.6 million years. The tundra marches down ahead of the glacial ice, the forests retreating below the tundra; the glaciers melt back, the tundra retreats back north with them, and the forests reassemble, with most (if not all) of the original cast of characters. Ten thousand years ago there was a one-half-mile-thick sheet of ice over New York State; today there are forests (where steel and concrete haven't replaced the woodlands, that is). And when those glaciers once again start their way south, as is widely expected to happen no more than

2,000 years from now, once again the habitat belts will move in a slow, but well-choreographed, dance.

Habitat tracking is the name of the game. Those that fail to track their habitats, find, as the most general rule, that extinction awaits them. Natural selection seldom—if ever—will take a species living in one place and modify that species as time goes on to meet the challenge of changing environmental conditions. Species get out, finding recognizable habitat as long as they can possibly do so. There is no reason to change under such circumstances, and species remain the same despite great changes in their distributions. G. R. Coope, a paleontologist and expert in Ice Age (Pleistocene) beetles, has documented wide changes in the distribution of some of his beetle species— species that are still alive today, but whose habitats were greatly disrupted as the European glaciers came and went. Yet the beetles appear much the same throughout the Pleistocene as they are today.

Extinction often ensues if no available habitat can be found. Extinction, the geological record shows clearly, has been going on all the time. The physical environment of the earth is *never* constant. Not finding suitable habitat, or experiencing competitive problems over habitat or resources with other species, leads, on a regular basis, to extinction. This is what David Jablonski, a paleobiologist at the University of Chicago, calls background extinction. Again, we get the feeling that extinction is not only real but a regular, usual thing—part of the regular cadence of life. But it is not, necessarily, a good thing.

It is the original Darwinian assumption that, faced with environmental change, evolutionary change will keep pace with changing times. Many biologists today still feel life is something of a horse race. The environment changes, and the holy grail of perfect adaptation to life's exigencies is a never-to-be-realized goal: Just as a species gets good at it, things change. A species changes, over the generations, through natural selection, only to find its goal receding. Species can never win. The result: A species existing through geological time keeps chang- .

ing. If this isn't exactly progress, at least it amounts to constant change.

But there is a major problem with this imagery. The fossil record (and this has been known to paleontologists ever since Darwin's day) clearly shows that, once they first show up, species usually don't change much at all. Species vary a bit as they go through time, but rarely do they show the kind of protracted change that some evolutionists, from Darwin to the present time, seem to think they *ought* to show. Natural selection is indeed a strong force. But, for the most part, it is a conservative one: As organisms chase suitable habitats around as the environment changes, they survive just fine pretty much in the state their ancestors were in originally. It is not change or die, but rather find suitable habitat or die.

So much for the link between evolution and improvement, or progress. Life seems finely tuned toward survival, and survival of a species generally entails many millions of years in the marine realm and hundreds of thousands to a few million years for terrestrial organisms. That survival of a species is achieved mostly through habitat recognition, and not through constant change as a means of staying fine-tuned to changing environments.

On the other hand, evolution—at least, appreciable evolution, true, noticeable changes in one or more of the adaptive properties of organisms—seems to occur mainly in response to new opportunity. The emerging picture is that the world's ecosystems, whatever and wherever they happen to be at any one particular time, are usually fairly full. New species occasionally evolve; some occasionally go extinct: background speciation and extinction. Nothing really grand is likely to happen.

New opportunities, though, do arise: When complex life first got going, there was opportunity galore. Nobody was there ahead of these life forms. Life had to invent ecosystems. This was also the case when life invaded the land for the first time (at the end of the Silurian). Radiations into a whole variety of environments by a number of different lineages were accomplished relatively quickly—once plants had established an economic beachhead, that is.

But what happens when ecosystems are entirely filled? Where are the opportunities? Extinction, by knocking out existing ecosystems, effectively sets back the evolutionary clock, though thankfully not to zero (life, fortunately, never became entirely extinct!). But it does allow new ecosystems to be formed, and new species to arise, at least in part in response to vacancies. The greater the extinction, the greater the opportunities, and the more evolutionary change will occur.

But back to our question: How do we reconcile this picture with the conservationist's concern that evolution is deterred by extinction removing genetic information? The answer—or so I firmly believe—comes from the mistake of thinking of evolution as a good thing. We should see life in a constant struggle not to evolve but to survive. Evolution is a fundamentally historical process: It has given us life's history (with a major role played by extinction); it will give us life's future, whatever form that happens to take. I think the message is fairly clear here: Life will have an evolutionary future; it will depend very much on what does and does not become extinct in the interim. But that is the future, and our concern really ought to be with the here and now.

The name of the game—the game of life as played at any moment—is not future evolution but survival. It always has been. We are sentient creatures, so there is every reason for us simply to realize consciously what we are indeed both expected and entitled to do unconsciously, just as every single organism does: *survive*. Ironically, if we manage to survive—and to help the components of all the world's ecosystems survive (which must happen if we ourselves are to make it)—*that* will throw a damper on evolution more than will our or any other species' extinction. We need to conserve genetic diversity—ours and other species'—to maintain the status quo, and not because of some imagined effect this will have on the evolutionary future.

Evolution has been my professional occupation for more

than twenty years; I see extinction as a fascinating spur to evolution—so much so that the modern biota, certainly including ourselves, would not have appeared unless past terrestrial ecosystem occupants had not been summarily erased. Had the dinosaurs survived the Cretaceous extinction, we mammals would still be scurrying around like the "rats of the Mesozoic world" that we were, as paleontologist Al Romer used to say. But once here, it is only natural that we want to avoid extinction—ours, and all of our fellow global ecosystem members.

Madagascar Now—and Then

The events frozen in the rocks at Stevn's Klint are remote. The evolution of life, punctuated with occasional severe bouts of extinction, intrinsically fascinating though many of us find it, would be of little import to humanity at large were it not for our increasing realization that we may very well be in for—or are already in the midst of—an episode of extinction. Past extinctions loom large as we grapple for a sense of understanding and a way out of our own environmental pickle.

A walk through a tropical rainforest is a far more immediate experience than a casual visit to Stevn's Klint is likely to be. At Stevn's Klint we experience chalk layers with only a few rather tough-to-extract fossils separated by a thin layer of clay whose ominous message must be wrested through laboratory analysis of its off-the-scale content of exotic rare metallic elements. A tropical rainforest, in stark contrast, yields a riot of life: tangles of vegetation, with hundreds of plant species represented within but a few acres. The cacophony of songs reveals the presence of many different species of birds, even though all birders know how frustrating it can be actually *seeing* their quarry in a rainforest setting. Life, in its exuberant abundance in the tropics, at first glance masks the deep trouble that most rainforests face these days.

Madagascar is a beautiful microcosm of the glories of tropical nature; it is also a fascinating laboratory of both evolution and extinction—a window of insight on what has occurred in the past and what is happening right now. Sitting astride the Tropic of Capricorn, most of Madagascar lies firmly within the tropics. The world's fourth largest island (about 1,000 miles long and 300 miles wide), the varied habitats of Madagascar support a rich array of life. Yet its animals and plants are strikingly unlike those of its nearest larger neighbor. At the narrowest point of the intervening Mozambique Channel, East Africa lies only 220 miles to the west of Madagascar. But Madagascar rifted apart from Africa at least 150 million years ago, and no one visiting Madagascar for the first time would mistake it for the Serengeti Plains of Tanzania. There are no lions or leopards, no mammalian carnivores at all even remotely approaching that size. The *fouche,* a civet cat about the size of a basset hound, takes top honors as Madagascar's biggest mammalian carnivore.

Large carnivores prey on large herbivores, and there simply are none in Madagascar: no giraffes, elephants, buffalo, or antelope. There was, until very recently, a species of pygmy hippo living there, a reminder of some affinity with African life. But that's it—no fossils of other familiar African animals turn up with the hippos; only the bones of living (and recently extinct) species of purely Madagascar animals show up. There has been extinction in the very recent past in Madagascar (more about that later), but for the most part the life of Madagascar's not-too-distant past is as un-African, purely Malagasy, as it is today.

Sometimes, where large mammalian carnivores are missing from an ecosystem, their place is taken by other sorts of vertebrates. The Komodo dragon (monitor lizard) is a huge nasty beast (thrilling in its fearsome ugliness, actually), that attacks and consumes deer, wild pigs, livestock, and (occasionally) humans on the Indonesian island of Komodo and its nearest neighbors. Huge pythons sometimes fill the same bill: Reticulated pythons are the largest carnivores on Sulawesi, specializing on the two species of wild pigs and other mammals (again, occasionally including humans). But, again, the absence of truly large mammals in general means that their carnivorous predators, whatever they may be, will not themselves be excessively large. There are plenty of snakes on Madagascar (none of them poisonous—shades of another island, Ireland), but relatively few reach as much as 6 feet in length.

While it is true in most parts of the world that the animals most dangerous to humans are other humans (probably as true in Madagascar as anywhere else), the nonhuman animals posing the greatest threat to human welfare in Madagascar are mosquitos, which presently are carrying a great many strains of malaria, many of which appear to be increasingly resistant to the conventional items in the medical arsenal used to combat the disease. (*There* is evolution in action, and hardly progress, at least to human eyes: Mosquitos carry organisms that cause the disease. Strains tolerant of quinine and other drugs are becoming more numerous each year through a simple process of natural selection.)

Yet Madagascar teems with both animal and plant life. Continental Africa has a single species of upside-down tree, the bizarre baobab, with its massively thick trunk and ridiculously skimpy branches, making the tree look as though it lives with its top buried in the soil and its roots projecting into the air. All the rest of the world's nine species of baobab are in Madagascar. Its mammalian fauna is also striking. There are no fewer than thirty species of the primitive insectivore genus *Tenrecs* scurrying about the forest floor.

But best known of all are the primates—all lemurs. These spectacular animals, superficially like long-snouted monkeys, lack some of the evolutionary features that monkeys share with us and the great apes, and so are considered to be rather more primitive members of the primate clan. Lemurs are only known from Madagascar, (one species has gotten to the nearby Comoro Islands). There are thirteen genera, twenty-four species, and many distinct divisions of some of these species, amounting to some forty-six different kinds of lemurs living there today. More are still being discovered in the forests, which is remarkable considering the relatively modest size of the island, and the interest it has attracted from naturalists for more than 200 years, as well as the fact that the Malagasy (as is generally true of local peoples the world over) are intimately familiar with their local wildlife. (The recently discovered golden bamboo lemur was known to locals in the rainforest near Ranomafana through its unusual vocalizations, but not assumed to represent an utterly distinct form of lemur.)

Lemurs provide a critical key to understanding how ecology and evolution fit together in Madagascar; extensive, if rather young, fossil deposits of their recent past help locate the current state of Madagascar's fight against extinction in the larger scheme of Ice Age extinction and have much to tell us, as well, of the human role in Malagasy extinction. We will be returning to them in the course of our coming to grips with what extinctions are and how past and present ecological crises relate to one another.

→ Madagascar is known for its strikingly high degree of _ende-mism._ An amazing percentage of the animals and plants on Madagascar are found nowhere else on earth. This is true, for example, of the lemurs: Four separate families (usually joined together as the superfamily Lemuroidea) are wholly unique to Madagascar. Eight of the nine baobab species occur there and nowhere else. Even birds, who can fly and would be expected, all other things being equal, to show up in other places as well, conform to this pattern to an amazing degree. There are some 250 species of birds so far reported from Madagascar. Of these, fully 150 are endemics: species that breed only in Madagascar. A few, such as the broad-billed roller, may migrate to East Africa in the nonbreeding season.

Bird endemism is not confined to species: There are several groups—families or subfamilies—that are unique to Madagascar: They evolved there and so far, at least, are wholly restricted to that island. The couas (an offshoot of the cuckoos) consist of nine species (though one appears to have recently become extinct). They are magnificent, rather large and puffy birds with blue patches around their eyes. There are the vangids, fourteen species, that recently have been linked to the shrikes, which are elsewhere so diverse, especially in the old world. Then there are the three mesite species—birds that basically no one knows how they fit into the grand scheme of evolutionary history.

Madagascar really has been an evolutionary laboratory. Because of its long, and obviously effective, isolation from Africa (and everywhere else), and because its modern fauna was derived from just a fraction of the different kinds of creatures that have been stocking the African ecosystems for millions of years, the modern fauna in Madagascar looks rather different from all the others. And because the ecological mix was unusual from the beginning, evolution has produced some interesting results. No fewer than four separate lineages of birds have become wholly, or at least mostly, ground-dwelling. There is a species of turkey-like guinea fowl that is shared with the African mainland. But the giant coua (and several other coua species) is a homegrown product, looking superficially like a pheasant. The ground-rollers likewise are unique to Madagascar, as are the aforementioned mesites, strange birds that are also secretive denizens of the forest or desert floor (depending on the species). Why so many different lineages of birds coming down out of the trees to risk life on the ground? Presumably because there are rather few terrestrial carnivores to pose much of a threat.

Life on Madagascar is unique; but life everywhere is unique because everywhere there are plants, animals, fungi, and micro-organisms that are found there and only there. Madagascar is simply more compellingly unique. It has a higher level of endemism, as the jargon goes. And, from our brief little visit, we can already see enough to know that life on Madagascar is unique in two rather different respects: Its animals and plants are, to a very great extent, restricted to that single island domain. To an evolutionary biologist or a conservationist this implies that a considerable portion of the earth's genetic diversity is locked up in species restricted to a rather small, and rather threatened, area. But Madagascar is also ecologically unique: The kinds of functioning ecosystems—habitats whose different populations are interacting in myriad ways, often involving energy exchange—are put together in rather unusual ways too.

Ecosystems are built from available components. They consist of parts of whatever species are available that can live in a particular environment. The structure of ecological communities depends on what evolution has produced, and what species have managed to survive to the present moment. Yet the game of life is played out, moment-by-moment, in the ecological arena—and the outcome of what goes on there determines, in no small way, which species indeed will survive and which will become extinct. The ecological arena also determines what form the surviving species will have in the future: What goes on in ecosystems determines the direction of natural selection that shapes future generations within each and every species. There is a tight connection, a give-and-take interaction, between ecology and evolution, and places like Madagascar help us zero in on the nature of that relationship.

We must worry that humans are at the root of much of the current threat to continued existence of many of the world's ecosystems, or many of the world's species (not the same thing, but both are legitimate ways of looking at the problem). But before we place the blame for all of the world's current ills squarely on our own shoulders, we should understand what factors are actually built into the workings of ecosystems that might also lead to extinction as well as evolution. Madagascar and other modern, vibrant (if threatened) ecosystems will help us. So will the fossil record of mass extinction and proliferation—the pulse of life's history on a grand scale, a history that (for complex organisms) has gone on willy-nilly without a human presence for nearly a half-billion years. We might be the root of many of today's evils, but extinction has been going on a long time without any help from us.

There is no doubt that Madagascar is in deep trouble—that is, insofar as the future of its wildlife is concerned. The excitement and joy engendered by that superficial first walk through a relatively pristine rainforest, with its riotous assemblage of plants and animals, are soon tempered by the hard-to-miss signs of trouble: Forest paths cleared the better to track lemur troops soon become overgrown with a tangle of invading, foreign plants. Far worse, the periphery of the forest everywhere seems to be ablaze: Farmers are clear-cutting and burning forest to secure new agricultural land. Part economic necessity, perhaps part more simply of cultural habit, even the steepest hillsides are being denuded of their trees as locals wrest a living at the expense of hundreds of other species trying to do the same thing. But, here again, Madagascar is but an example of a problem pandemic throughout the world.

We of the developed nations have far outstripped our fellow humans in the tropics in destroying primordial habitat for the very necessary and wholly understandable purpose of feeding ourselves. That residents of tropical countries are very much aware that our moralizing is more than a little presumptuous and hypocritical should come as no great surprise. These are the very real and fascinating politics of land use and conservation, and once again Madagascar serves as an entrée, a microcosm, to this most crucial of arenas, one that we will explore in greater detail in the final chapter.

But first we need to understand more about how nature is put together, how it runs, how its history is entwined with its present structure. We need to know more about how that system can break down and rebuild. We need to know more about how mass extinction can occur without any help from us before we can assess our own role in extinctions of the immediate past, and especially the present and near future. For that we need to take a closer look at what the term *diversity* really means.

CHAPTER TWO

Santa Rosalia, or Why Are There So Many Kinds of Living Things?

IN 1959, ecologist G. E. Hutchinson published a brief paper in the technical journal *American Naturalist*. Entitled "Homage to Santa Rosalia, or Why Are There So Many Different Species?," Hutchinson returned to an age-old question fundamental to understanding life on this planet. Why, indeed, are there so many different kinds of plants, animals, fungi, and microorganisms? Have these numbers always been about the same? What determines how many species can be represented in a single local habitat or ecosystem? Are the factors that determine the numbers of species within a basic taxonomic division, such as the number of species of American thrushes, connected in any way to the factors that limit the number of different thrushes in a given habitat? Answers to these questions matter deeply to our quest to understand what happens, what actually goes wrong, when ecosystems collapse in a paroxysm of extinction.

No one has a truly solid idea of how many species of organisms actually inhabit the planet right now. Indeed, one reason why there is such urgency to conserve tropical terrestrial habitats is that species of plants and animals are typically confined to much smaller areas in the tropics than in the higher latitudes. Cutting and burning a few hectares in the Amazonian rainforest threatens entire species—species that, in all probability, have yet to be sampled and studied carefully. We risk losing species before we have even recorded their existence. This is especially true of insects, which exist in numerical profusion as individuals, and are packaged, as well, into countless millions of species.

When I was in graduate school in the 1960s, the standard estimate of the current number of species on earth was 1 million. In the early 1980s, Smithsonian entomologist Terry Erwin sampled the insect faunas of the treetops of a small area in Brazil. He found such a high level of endemism (restriction of species to but a few hectares), and such a high number of previously unknown species, that he decided the true number of species alive today must be about 30 million. More recent estimates put the figure upward of 80 million. We can only assume the actual number lies somewhere in between, and admit, with such disparate high and low estimates, that we really have no precisely accurate estimate of present-day diversity at all.

The job of tabulating life's diversity falls to experts in particular groups of organisms—taxonomists, or systematists as they generally prefer to be known. Systematists classify organisms into species, naming new species whenever samples turn up that can't be recognized as belonging to a species already known. Beyond that time-consuming and all-important task, systematists have the further goal of determining how all these species are related to one another: Their job boils down to figuring out the evolutionary relationships linking all the diverse species. It is systematists (paleontologists as well as those working exclusively on modern creatures) who decipher the evolutionary history of life. As a paleontologist, I have discovered and named many species of extinct trilobites and horseshoe crabs. But the work does not stop there. I have analyzed the evolutionary relationships between species, which has

22

led to new insights into the very nature of the evolutionary process.

One fruit of these pursuits is the extended classification we give each organism—a direct reflection of its evolutionary affinities: Thus every organism, say a fruit fly, belongs to a species; *Drosophila melanogaster* is but one of many species belonging to the genus *Drosophila*. These species of closely related genera form the Family Drosophilidae of the Order Diptera ("two wings") of the Class Hexapoda (or Insecta) of the phylum Arthropoda (which includes crustaceans and a number of other extinct and recent organisms). Arthropods belong to the Kingdom Animalia. Evolutionary history fairly reeks from a simple recital of the classification of any creature. And, it goes without saying, we need such monickers simply to be able to speak about these creatures: to count up how many of what sort are around, to compare these numbers with similar counts in past times, or in other places, and to evaluate just what sort of threat of extinction—and for what reasons—these creatures face.

Systematics is an old and honorable profession. In the nineteenth century, when nothing much was known of the world's organismic populations, systematics was at the very forefront of science. Back then, descriptions of new species were routinely included in the annals of the *American Association for the Advancement of Science*. That it has been to some extent eclipsed, repeatedly, as biological science takes advantage of ever-more sophisticated technologies, probing the ever-smaller physical and chemical realms within cells as its new frontiers, is by no means to disparage the continuing importance of systematics in the modern world. We simply cannot evaluate how the biota is changing—disappearing—without a clear and accurate vision of what actually exists out there. Systematists alone can tell what species have become extinct, just as it is they who tell us if a newly collected specimen represents a species no one has ever seen before.

Most systematists these days work in natural history museums, though there are a few, in ever-decreasing numbers, still to be found on university campuses or in governmental agencies. Systematists rely heavily on extensive collections: The basis of their work is comparative. After all, the only way you can tell if you have something new, or determine to what other species a certain species is related, is by carefully comparing specimens. The great natural history museums of the world, most of them founded in the mid-nineteenth century coincident with the latter stages of European (and later North American) colonial activity, are fantastic treasure troves of the world's past and present biological diversity. They are nothing short of national treasures: records of the world's diversity as it was at the beginning of the nineteenth century. Museum collections give us the baseline for interpreting what is going on around us right now in the living world.

Yet G. E. Hutchinson is an ecologist, not a systematist (though he has, indeed, made contributions to systematics throughout his long and distinguished career at Yale). Ecologists look at the world in a way that cuts right across the systematist's perspective. Systematists, specializing, say, in beetles, will go to a spot and collect a sample of all the beetle species present. Sometimes they'll collect more widely, bringing back other sorts of insects as well, for colleagues down the hall working in the same department. But beetle systematists generally won't look at the birds, earthworms, mammals, and mollusks also there, and, with the possible exception of the actual plants on which the beetles are living, little notice will be given to the vegetation. A beetle systematist is interested in *beetles* at this one spot, and elsewhere as well; the same species probably occur elsewhere, and their nearest relatives are likely to occur not too far away; evolutionary history embraces the entire geographic range of a single, evolutionarily pure group of organisms.

This is hardly of moment to an ecologist. An ecologist is interested in *everything* present in a single habitat. Beetles are eating something, and, in turn, most likely being eaten as well. Ecologists want to know how many different kinds of organisms are present in a particular habitat. They want to know what the relative numbers of each kind of creature are. They are concerned with the stability of the relative predominance of creatures as time goes by. Ecologists want to know why there are so many species of all sorts of different kinds of organisms represented in the local system. Systematists want to know why there are so many species of fruit fly in the entire world. The perspectives of these two sorts of biologists are intriguingly different.

As we shall see, we will need to integrate elements of *both* perspectives to frame a comprehensive theory of extinction. Each group speaks of diversity, seeks to understand what determines levels of diversity and what causes fluctuations in diversity: increases in numbers that either enrich ecosystems or add (sometimes explosively) to the numbers of species in taxonomic groups; or fluctuations that decrease numbers, threatening ecosystems, or driving large-scale taxonomic groups, such as the Family Elephantidae, to the brink of extinction. (There are but two species of elephants left, compared to the minimum of ten species known to have existed during the last 100,000 years.)

Patterns: Land vs. Sea, Tropics vs. Higher Latitudes

It has been obvious, in a general sort of way, for centuries that there are more different kinds of animals and plants the closer to the tropics one travels. European naturalists were aware of this pattern, as indeed were savants of the ancient Mediterranean before them. Anyone living in the more northerly climes of North America immediately becomes aware of the greater array of insects, flowering plants, and perhaps even birds as one travels south, especially if a vacation to Hawaii should ever materialize. Naturally enough, people have tended to wonder why there are more different kinds of plants and animals in the lower latitudes.

So the problem is not just why there are so many plants and animals on earth but why they are concentrated more in some areas than others. In the seas, which currently cover 71 percent of the earth's surface (a lower than usual figure looking back over the past 600 million years), we find the same pattern from high to low latitudes: There are far more marine species in the tropics than we see as we sample progressively closer to the north and south polar regions. Yet there is another contrast: One would imagine that, with the seas being so much more extensive, providing vastly more habitat space than terrestrial environments, there would be proportionately far more species in the oceans than on land.

However, there are far more terrestrial than marine species, though it must be admitted that it is far easier for us terrestrial humans to encounter organisms on land than within ocean muds, or on the lightless sea bottom under a mile or so of water. The recently discovered deep-sea vent faunas, where sulphur-fixing bacteria form the base of a food chain whose energy is derived from hydrothermal vents (hot springs) rather than sunlight, shows how little we have yet found out about life's diversity in such hard-to-reach places. Nonetheless, comparison of subaerial with aqueous near-shore communities shows up a pattern borne out by sampling in all oceanic depths so far: There simply are more different kinds of animals and plants on land than in the sea.

We need to know why these patterns exist. What causes systematic variations in diversity patterns, what regulates and controls species numbers by region, by ecosystems, and by organismic group is crucial to an understanding of what really causes decline in diversity, and how likely it may be that such declines can be reversed. Should we try to save species or habitats? There is an urgent and practical significance to what might otherwise seem to be rather abstract and esoteric biological theory.

One common thread that links the two comparisons—tropics vs. higher latitudes, land vs. sea—is habitat patchiness. The contrast is more vivid and intuitively obvious when we think of the vast, monotonous stretches of sea bottom in contrast with the typical choppiness of the terrestrial realm. Hills separated by valleys with streams emptying into, and running out of, lakes dotting the landscape offer a much more heterogeneous picture than endless vistas of sandy bottoms below the waves that are out of reach of disruption by even the mightiest of storms, an environment with little year-in, year-out variation in temperature, salinity, or amount of oxygen dissolved in the seawater.

To be sure, there are monotonous stretches of environment aplenty on land. True deserts may bloom briefly with a beautiful array of flowering plants, and the wet season may support for a while herds of nomadic antelope and other grazers, along with their respective predators, but relatively speaking, deserts are biological wastelands of rather low diversity. Conversely, coral reefs (confined to the tropics) are intricately heterogeneous. The colonial corals (and sponges and other, rather simple, invertebrate colonial organisms that form the actual building blocks of reefs) form massive, yet highly irregular, structures that jut up and away from the seafloor. The nooks and crannies of such reef systems offer haven to a profusion of organisms of different species, supporting, in turn, a rich array of fishes living just off the face of the reef, deriving their livelihoods from the creatures living in or on, or actually forming, the reef itself.

But these are exceptions that prove the rule: The more physically heterogeneous an environment, the more different sorts of organisms will tend to be found there. This is true for the land and for the sea, though, as we have seen in general, there is less heterogeneity in the sea than on land. Environmental patchiness, or degree of heterogeneity, is traditionally significant in both ecological and evolutionary explanations for systematically varying patterns of diversity. And herein lies yet another fundamental distinction between the concerns of evolutionary biologists and ecologists: Ecologists want to know why more different kinds of species are represented in some environments (habitats) than in others. Theirs is a question of how ecosystems function, and how diversity patterns are consequently maintained.

Species come and go, under normal circumstances, within local habitats. Hairy woodpeckers may be present for several years in a woodlot, disappear for several years, only to come back again in due course—just to cite the backyard habitat I know best. Populations crash and are replenished from populations elsewhere. Ecosystems are not machines that function only when absolutely every component, every organismic cog in the wheel, is in place at all times. Communities of different sorts of organisms fluctuate, often considerably, in composition from season to season and year to year. The parts of the system—the local populations of various different species—may come and go, becoming locally extinct or simply migrating away. The species continues to exist elsewhere, and will, as a rule, in due course be represented sometime again in that local ecosystem from which it disappeared.

Evolutionary biologists, especially systematists, do not look at the world in quite that way. Systematists tend to be much more historically minded people: They want to know how particular species evolved. Evolutionary biologists in particular tend to focus on the adaptations of organisms: those features of anatomy, behavior, and physiology that fit the organisms to their environment, allowing them to obtain energy resources simply to grow up and live, to avoid predators and withstand climatic rigors, and also to reproduce. How did hairy and downy woodpeckers, with their different beak sizes adapted for slightly different feeding behaviors, evolve in the first place? An ecologist asks: Why are hairy and downy woodpeckers here right now, in this particular ecosystem with this particular array of other species?

An evolutionary biologist wants to know how it was that

these two particular species evolved in the first place and managed to survive, and what shaped the particular, and often unique, battery of organismic adaptations that characterizes the two species. The ecologist is interested in the here and now, with scant attention to history: It is the moment-by-moment dynamics of the system that seem to offer the answers to diversity questions in an ecologist's mind. Evolutionary biologists tend to take the longer view, explaining the configuration of organisms, including the explanation for diversity questions, as an outcome of a long-term, historical, evolutionary process. But, in so doing, an evolutionist typically overlooks the fine-scale interactions that are the very stuff of ecological dynamics.

Different sorts of questions, different kinds of answers, and yet, somehow, the concerns of ecologists and evolutionary biologists must surely be related. After all, an ecologist might well say "There are so many different species living here because each has its own niche"—its own specific economic role to play in the ecosystem. An evolutionist, on the other hand, is concerned, at least in part, with explaining the origin and maintenance of organismic adaptations—the very features that, when put to use, perform those very ecological roles an ecologist has in mind when speaking of niches.

Evolutionists surely must have function in mind when they explain modification of adaptations by natural selection through time: All modern horses are grazers, with high teeth loaded with tough dental "cement," the better to resist wear from the highly abrasive siliceous grasses that appeared 20 million years or so ago, back in the Miocene. (Virtually all of us have been cut by aptly named "blades" of grass, which slice so well because they are loaded with tiny particles of glass.) When paleontologist George Gaylord Simpson, young but soon-to-be-renowned evolutionist and lover of horses, analyzed equine evolutionary history, he focused heavily on the transition from browsing on leaves to grazing on grass: a major element of horse evolution, a switch from one basic mode of feeding to another, which underlay much of the anatomical change in Simpson's fossil horses.

Simpson used horse evolution, particularly the relatively rapid switch from browsing to grazing in one lineage (with some browsers living on for a while before finally succumbing to extinction), as the centerpiece example of his celebrated theory of quantum evolution. As developed in his 1944 book, *Tempo and Mode in Evolution,* quantum evolution (which persists in modified form in evolutionary theory to this moment) explains major events of evolutionary change as rapid shifts in the adaptive properties of organisms within lineages. Paramount in all this is a consideration of the functional aspects—closely akin to the ecological roles—of the parts of organisms that are undergoing evolutionary change.

So ecological niches—the roles that organisms of various different species play in local ecosystems—are central to both ecological and evolutionary explanation. Extinction most certainly includes ecosystem collapse, spelling niche loss, which can lead to the demise of entire species. Natural selection forms the usual evolutionary account of the origin of the adaptations that form the basis of an organism's niche activities. Natural selection reflects what works best in the parental generation: Those organisms most successful will tend to leave more copies of their genes to the next generation (simply by producing more offspring) than less successful organisms of the same species. *Success* here means relative ability to make it in a (generally) tough world: to withstand climatic conditions, avoid predation (being eaten), and, above all, secure an adequate supply of food—the quintessential energy resources needed simply to live.

And, clearly, these activities all go on within an ecological context: within communities in a local habitat. In other words, part of any organism's environment is all the other organisms—of one's own species and, perhaps more importantly, of all other species represented in the local ecosystem. Natural selection tunes adaptations to the characteristics of organisms of other species, as well as to the conditions of the local physical environment. Here is where ecology and evolutionary biology, truly do meet: the notion of *coevolution* (still something of a hot topic in modern biology) looks at the close association, even interdependence, of organisms of different species—say, a butterfly dependent on a single plant species as a food item. The plant, in turn, relies completely on the butterfly for pollination. Both species need each other. Their fates are intimately intertwined. More generally, in less extreme cases, the adaptations of organisms of any given species are constantly being affected, each generation, by changes in other species that are also parts of the local ecosystem.

But what does all this have to do with the patchiness of habitats that seems so closely tied in with differences in patterns of diversity? Simple—at least as a first-order generalization: The greater the environmental heterogeneity, the greater the number of different microenvironments there are likely to be. The chances are greater for a wider variety of organisms with different basic physiological requirements to congregate. And the more different kinds of organisms, the more *additional* kinds of organisms will be able to fit in: more different predators, for example, the greater array of herbivores, which number is itself in turn sensitive to the variety of different kinds of plants present. Complexity begets additional complexity, but rests, at base, on environmental heterogeneity—the bottom line (perhaps *prime mover* would be better) in determining how many niches in fact will actually be represented in a given ecosystem.

We are almost in a position to see how ecological and evolutionary explanations of their respective types of diversity patterns actually converge. We need but one additional item, again pertaining to the notion of ecological niche. And that is simply the idea that niches have "widths." Horseshoe crabs are invariably the last of the larger organisms to be ousted from polluted estuaries along the eastern North American coast. And that's simply because they can take a lot. They are tolerant of wide ranges of salinity and temperature. They can even survive long stretches out of water, exposed to direct sunlight (horseshoe crabs emerge to mate and lay eggs on sandy beaches, so they are no strangers to the subaerial environment; several of their extinct relatives in the Paleozoic apparently became exclusively terrestrial, as have their still-living kin, scorpions and spiders). To top it off, horseshoe crabs eat a wide variety of different organisms, including small clams of various species and marine worms of various sorts. They are omnivores, accepting whatever they can grab along the sea bottom.

Horseshoe crabs are a great example of a broad-niched species. In ecological parlance, they are eurytopes. Narrow-niched species (stenotopes), in contrast, are specialists. They may concentrate on but a single food resource, such as the butterfly dining off only one single species of plant. They may tolerate only a narrow range of temperature, or salinity if they are aquatic. Clearly, the terms *eurytopic* and *stenotopic* are relative, and, as a matter of fact, best used comparatively only when talking literally about close relatives. There are three additional species of horseshoe crabs (all living in eastern Asian waters), and, as far as is known, they too are relatively broad-niched. But interesting contrasts emerge when closely related species are compared, as we shall shortly see in two closely related groups of African antelope.

Niche width can be a difficult, even deceptive, notion to apply to nature. Without two species to compare, what one person might take to be a fairly broad-niched species could well be another's example of a perfectly good narrow-niched form. Worse, organisms within a species might be broad-niched in terms of food resources, but narrow-niched in some other respect, such as temperature tolerance. Some organisms undergo radical changes in their ecology as they grow up to become adults. (Human children have, as their main job, understanding their parents sufficiently to ensure continued support—a situation that generally, and inevitably, changes as the years roll by.)

More germane to the point of changing niche widths through different stages of the life cycle, consider the effect that mere size has on increasing the array of available food resources as a typical predator gets older. Sometimes, entire ecologies change: Young Komodo dragons tend to hunt in trees, taking small prey and, incidentally, avoiding being eaten by the older members of their same species. Older Komodo dragons eat deer, pigs, domestic livestock and, occasionally, tourists. Usually, though, the repertoire of prey items simply expands as predators reach full adult size.

Yet, however tricky the notion of niche width may prove to be, it has long loomed as an inviting source of understanding those vexing differences in diversity ecologists have been pointing to for years, especially between the tropics and higher latitudes. Put together with notions of relative habitat patchiness, niche width considerations go a long way toward telling us about the dynamics that control diversity patterns in the ecological arena and in the evolutionary realm as well. Let's take a closer look to see how this might all fit together to explain both ecological and evolutionary diversity patterns, and what the common threads that link the two different sorts of systems might really be. What generates diversity patterns in the first place has much to tell us about how those patterns fall apart in extinction.

The Living Dead and Ecosystem Diversity

Conservationists these days are especially worried about the fate of tropical ecosystems. At first glance, this may seem puzzling, as the tropics have far more species than the higher latitudes: One might be tempted to conclude that the tropics would be less, rather than more, vulnerable to extinction. Instead, biologists see tropical ecosystems as much more fragile and extinction-prone than most ecosystems of the higher latitudes. We need to understand why.

Why the tropics should have more species—different kinds of organisms—than higher latitudes is, as we have already seen, a long-standing puzzle. Coral expert Brian Rosen has pointed out that explanations for the phenomenon fall essentially into two distinct categories, perfectly reflecting the distinction between ecological and evolutionary explanations. Some biologists see the problem in purely historical terms: The tropics are richer in species than the higher latitudes perhaps because they have acted as a pump, a source of a higher rate of production of new species than in colder climes. Or, again within the camp of evolutionary explanation, it might simply be that extinction goes on more slowly in the tropics: Tropical areas may be a sort of living, outdoor, in-the-wild museum, accumulating more and more species as time goes by, perhaps simply because life is easier, more lush in the tropics than elsewhere. Some historically minded biologists, not surprisingly, see both sorts of processes acting in conjunction: Species appear at a faster rate and disappear at a slower rate in the tropics than in the higher latitudes.

A common denominator to all the variant versions of historical explanation is the constancy of relatively high amounts of available energy in the tropical regime. Organisms constantly require energy resources simply to survive; all but a few derive their energy ultimately from the sun—by far the most abundant source of energy on the earth's surface. The sun is an *exogenous* energy source. Plants, through photosynthesis, trap a fraction of the sun's daily energy yield, thereby providing the basis of all animal life.

Only a few kinds of microorganisms have metabolic pathways that allow them to live off the earth's *endogenous* energy source: the heat that flows from deep within the earth, heat that derives from radioactive decay. The earth's own energy is constantly flowing up through the mantle and crust; anyone walking down a mine shaft realizes that, after an initial chill just below the surface, it gets hotter and hotter as you go lower. The plates of the earth's crust collide, pull apart, or slide along past one another, driven by this endogenous energy. Earthquakes and volcanoes release this energy in spurts. And on the floor of the oceanic abyss, thermophilic bacteria trap that energy and serve the same role that photosynthesizing plants do in the vast bulk of the world's ecosystems: They serve as the base of the food chain, providing the ultimate source of available energy for all the rest of the creatures in the local ecosystem. Heat rising through the vents causes hydrogen sulfide to form, and bacterial chemosynthesis derives energy through the oxidation of the sulfide.

The tropics have been there for at least as long as complex life has been around—a good 600 million years. Maps of ancient geographies can, at first glance, be a bit confusing. A map of the world as it was 380 million years ago (smack in the middle of the Devonian geological period), for example, shows the equator running up catty-corner through North America, entering somewhere in the vicinity of modern-day Baja California, and exiting somewhere up around Hudson's Bay or Newfoundland. But the map does not signify that the tropics have shifted. Rather, North America, and all the other continental (and oceanic, for that matter) chunks of the earth's crust (tectonic plates) have themselves been shifting around.

North America was simply tilted on its side and lying astride the equator 380 million years ago. The equator then was where it always has been: midway between the poles. The geographic poles appear to have remained fairly stable, even though there is some evidence that the magnetic poles have done some shifting around themselves through the ages. And North America, blessed with a profusion of fossils from the shallow seaways that flooded the continent in those days, clearly shows that life was indeed tropical back then: Coral reefs were common in Devonian North America. And diversity was higher in fossil communities in North America along the ancient track of the equator than it was in Devonian communities in higher latitudes (colder habitats). The pattern of higher tropical diversity is indeed an ancient one, even though all the species alive in the Devonian have long since become extinct, replaced by a wholly new cast of characters.

Thus time and higher amounts of available energy seem to work hand in hand in generating and accumulating higher diversity in the tropics. Yet this can't be the whole story, because there is also the ecological perspective to be considered: How is it that more different kinds of organisms seem to be able to live side by side, integrated into single ecosystems? There must be some functional reason that the tropics can support more different life forms than higher latitudes—in the sea as well as on land. And this takes us right back to the notion of niche, and the intriguing possibilities offered by that spectrum of niche width.

Ecologist George Stevens has recently thrown a new ingredient into the old debate over diversity patterns in the tropics vs. the higher latitudes. His is a lateral approach, a way of sneaking up on the old question of higher diversity in the tropics by starting off looking at a different diversity pattern— one that he claims sheds light on the tropics problem.

Stevens calls the second pattern Rapoport's rule, after the biologist who was apparently the first to draw attention to the phenomenon. In group after group, there seems to be a gradation in size of geographic range of species: The higher the latitude, the greater the average area and latitudinal extent of a species' range will tend to be. Stevens has documented this pattern in plants and animals, vertebrates and invertebrates, and in both terrestrial and marine environments. The only exceptions, he contends, are the ones that actually prove the rule: Those organisms that migrate (such as many birds) or act like migrants (such as insects that are dormant through extended periods of the year) don't develop the pattern, simply because the ecological mechanisms that produce the pattern are circumvented by such migratory (or pseudomigratory) behavior.

Stevens thinks there is a link between the fact that there are relatively fewer species in high latitudes and Rapoport's rule:

that those relatively few species in the far north (e.g., tundra) have typically vastly more extensive ranges than tropical species characteristically display. The connection, as I see it, comes through the notion of niche width: We tend to think of the climate in the far north as harsher than in the tropics. But we think so probably only because we are ourselves a species of relatively recent, and almost assuredly tropical, origin. Cold-adapted organisms loathe heat the way most of us dislike extreme cold. What is different between the tropics and cold climes is the extremes—especially in temperature and perhaps in rainfall—that higher-latitude organisms face; temperatures vary widely over a year in the higher latitudes. High-latitude organisms must be able to take these extremes: They must be eurytopic, at least in terms of the climatic oscillations they routinely face. Eurytopic organisms, typically, can spread out over larger areas, recognizing suitable habitat over a broader geographic range simply because they can tolerate a relatively greater range of different conditions.

Ecologists have long realized that there is little point for tropical species to display an equivalently broad range of tolerances to climatic conditions. Why should they? The average midday temperature in Indonesia varies between 87°F and 88°F throughout the year. It is difficult to see how natural selection could shape organisms to have broad tolerances when in fact they are never exposed to widely oscillating climatic conditions. In any case, such a wide range of physiological tolerance would, in a very real sense, be an unnecessary physiological drain on a tropical organism. Freed from the necessity of having to cope with varying extremes, tropical organisms have instead focused closely on microhabitats: They have become narrow-niched specialists (stenotopes) whereas higher-latitude species tend to remain eurytopes, forced to be broad-niched by the very climatic circumstances in which they live.

There are several ways to look at the implications that niche width would have for producing the higher diversity we see in the tropics than in the higher latitudes. Stevens takes a rather novel tack with his use of famed ecologist Daniel Janzen's notion of the "living dead." An ecological fact of life too long ignored is that all species have limits to their distributions— limits that arise simply because a species will spread throughout a region wherever there is suitable habitat. The boundaries of a species' range reflect the limits of distribution of suitable habitat. Ranges can, and do change, however: Everyone has heard that the Africanized "killer" honeybees are steadily moving northward. Tufted titmice are increasing their presence in the northeastern United States. Titmice, along with a few other bird species, may well be able to tolerate northern winters because so many people stock feeders, especially during winter months. But other bird species are also moving northward: blue-gray gnatcatchers, a migratory (and nonfeeder participant) species, is breeding farther north than it did in the nineteenth century—more a reflection of the warming trend, presumably, than of any other factor.

Yet, at any one moment, the limits of a species range really are a reflection of current limits of tolerable habitat. There is a game often played as one drives south, descending from the Colorado plateau toward Phoenix along Arizona Interstate 17: The object is to be the first one in the car to spot a saguaro cactus—the archetypal symbol of the arid American southwest. Saguaros simply cannot live atop the Colorado plateau. In contrast, they are all over the place on the flat arid lands around Phoenix. Obviously, there must be a transition zone between total absence and profligate presence of saguaros along I-17. The first saguaros are loners, single plants in the lee of south-facing draws, up little ravines. A few plants manage to make it, eking out an existence, though a lonely one, a fact vital to Stevens' explanation of tropical diversity. Continuing south, down the escarpment, the saguaros begin to show up in other areas of the ravines, including, eventually, north-facing slopes. And they are more abundant, with several living in close proximity, until, finally, they are all over the place.

George Stevens would call those outpost saguaros the "living dead." Organisms living at the very extremes of a species range are rarely the healthiest, most vibrant members of their species; after all, the organisms living closer to the center of a species' range will be experiencing the more optimal conditions, and should, all things being equal, be having an easier time of it than those outposts at the very margins, living in the very extremes tolerable to members of that species. Typically stunted, they are also typically alone, or nearly so: Population densities, as with the saguaros, are far lower at the extremes than smack in the middle of a species' range. A consequence is that reproduction is a luxury that is sharply curtailed in those organisms living at the very edge of a species' range. They are the living dead because, though they can hang on and eke out some form of an economic existence, they themselves seldom reproduce. Their lives, in a very real ecological and evolutionary sense, are typically incomplete.

Such outposts—the living dead—get there as emissaries: as wind-blown or animal-carried seeds, or as wandering animal migrants. They are replenished not so much by their own reproduction as by continuing migration, the grabbing of an ecological foothold in meager conditions. Stevens thinks that one reason why the tropics have so many more different species than more northerly climes is simply that tropical ecosystems are filled with living dead. Because tropical species need not be generalists, but rather specialize, they recognize many subtly different habitats in the tropics and tend to be restricted to one particular specialized habitat.

Tropical ecosystems are typically rich in numbers of species, but poor in numbers of organisms—the exact opposite of the high latitudes, where there may be but a few species, each with astronomically large population sizes. Recall the common tendency in tropical ecosystems (such as those in Madagascar), for there to be well over a hundred species of trees represented in but a few hectares of land, yet many of those species are actually represented by only a few (often only a single) organisms.

Stevens thinks such situations may well be like those

saguaros on the edge of that species' range: effectively, "living dead"—hard to imagine in the midst of such tropical living exuberance, but perhaps the truth in any case. With the great array of microhabitats in the tropics, Stevens simply believes that dispersals may manage to gain a foothold from time to time in suboptimal habitat, depressing both their chances of flourishing and especially diminishing their chances to reproduce. Sex requires two, and realistically, it takes a local population of a lot more than two to be able to maintain itself—a fact all too familiar to conservation biologists trying to assess minimum numbers that must be present for a local population of a species to be able to survive by replenishing itself each generation.

But this fact takes us right back to habitat heterogeneity, and ultimately to the numbers of niches present. For regardless of how many of the species represented in a local habitat in the tropics might actually be living dead, they wouldn't be there if the main body of their species were not living somewhere not too far away. The living dead do not explain the ultimate differences in broad terms between the tropics and the higher latitudes; they just explain why more different species can be represented in a relatively small area in the tropics than in the higher latitudes.

It is, rather, the very existence of the microhabitats, each with conditions more suitable to the needs of some species than others, that ultimately supplies the functional side of the diversity argument. More different habitats, thus more different niches, reflect at base two things: greater microhabitat heterogeneity and the freedom that tropical organisms have from the burden of having to remain ecological generalists, allowing them instead to specialize, to focus on and take advantage of that very heterogeneity. Generalists (such as the higher-latitude organisms are) simply would not detect that heterogeneity because their physiologies would allow them to live in all those different microhabitats.

We now have a handle on why there are more different

species in the tropics than in the higher latitudes. There are more species and families of birds in the tropics than in the higher latitudes: the perspective of a systematist. But there are also more species of all manner of living things per hectare in the tropics than elsewhere, and this is an ecological pattern. How these patterns and perspectives converge through the concept of niche width, yielding further clues to the puzzle of extinction, will become clearer as we take a close look at a real-life example posed by African antelopes.

Antelopes and Niches

It's hard at first for an American on a casual visit to the great parks and game reserves in Africa to shake off the feeling of being in a zoo. Even if the perimeter of these vast spaces is fenced, it is really the animals themselves—especially, of course, the large mammals—that foster the impression. It can take days for the obvious truth to sink in: The animals are in the wild, in their accustomed habitat. The visitor is the one who is trapped in an alien setting, the car or tourist van confined to the tracks that wind through the preserve.

Yet there is something to this image of Africa-as-zoo that transcends the mere psychic misapprehensions of a foreigner reared on exotic displays in zoological gardens. Africa really is a preserve in a deep evolutionary sense: We see a zebra, or a lion, and we think of Africa. True, India and Indonesia have their share of rhinos and elephants, and tigers are equally as impressive as African lions. But Africa has so many large mammals: lots of different kinds and often (or until rather recently) many individuals within a species, as anyone who has seen herds of African buffalo will attest.

What makes Africa a zoo is the simple, if perennially shocking, fact that all these different sorts of beasts—elephants, rhinos, lions, hippos, antelope, and so on—covered much of the rest of the world as recently as 8,000 years ago. Think of it:

8000 years. That's *thousand*—still a lot of time in purely human terms, but literally next to nothing when we think of the hundreds of millions of years of life's history. Extinctions of modern species have been going on for many thousands of years now; one way to look at our current dilemma is to see ourselves simply in the midst of an ongoing major extinction event.

The great extinctions 8,000 to 10,000 years ago that have marked the end of the Ice Age are especially critical, as here for the first time the activities of our own species have been implicated: Hunters, precipitating the great Pleistocene "overkill," are assumed by many to lie at the root of the disappearance of many of these grand (literally, mammoth) mammals. Yet, because we know that many species, including beasts far larger even than elephants, have likewise become extinct at an earlier time without the helping hand of humans, the dynamics of Ice Age extinction should have much to tell us about extinction in general, and the exact nature of our species' role in it. It is definitely a hybrid sort of situation between the old mass extinctions that occurred perfectly well (if that is the right expression) without our help and the situation we face now, which is interpreted (perhaps not totally correctly) to reflect solely the misdeeds of our species inflicting change—extinction—on so many of our kindred species. (I will have a lot more to say about Pleistocene animals and plants when we try to come to grips with what really is going on in large-scale extinction events in later chapters.)

But Africa has, at least so far, resisted the large-scale end-of-the-Ice-Age destruction of its fauna and flora that has hit most of the rest of the world. The fauna is not strange, just lucky. It is, literally, a remnant, a holdover, of a fauna fairly typical of much of the world's surface. Not surprisingly, Africa affords a richness in both ecological and evolutionary insight. It is the double birthplace of our own lineage both in remote times, as we split off from the great ape lineage some 4 or 5 million years ago, and more recently, when our own particular species, *Homo sapiens,* emerged from Africa to spread throughout Asia, Europe, and eventually most of the rest of the earth's surface.

Africa has also posed some classic evolutionary puzzles: Why, for instance, are there so many different species of closely related cichlid fishes in each of the great lakes in the East African rift valley system? Cichlids are a diverse array of freshwater fishes found the world over, primarily in the tropics. There are perhaps as many as 1,000 cichlid species in Lake Malawi alone. How did they all become reproductively isolated from each other—in seeming contradiction to the maxim that geographic isolation is a necessary forerunner to the formation of new species? Interestingly, it is the by-now familiar notion that populations of different species have different characteristic niche widths that has underlain the most promising approach to solving the cichlid diversity question: Narrow-niched, relatively ecologically specialized species (such as mouth-brooders of the genus *Haplochromis*) tend to accumulate in larger numbers within local ecosystems, presumably by seeing greater patchiness in the environment than can their less specialized, more broad-niched kin (in the case of the cichlids, species of the genus *Tilapia,* widely introduced around the world as a fisheries resource, a reflection of the ecologically general nature of these fishes).

Perhaps the clearest example of the relation between niche width and the rate of both the appearance and extinction of species comes from yet another group of characteristically African animals—the antelopes. Africa is truly rich in antelope species, with sixty-eight of them still to be found in a wide range of habitats. Waterbuck and antelopes called lechwes prefer marshes and areas along watercourses; various species of oryx prefer the most arid of desert conditions; gazelles and other species prefer open grasslands; others like a mixture of grassland and woodland; and still others prefer the dense forests.

George Stevens, echoing a familiar ecological theme, noted the propensity for ecological specialists to occur in the tropics, with the more broad-niched species occurring in higher latitudes with their greater ranges of environmental conditions. Yet another pattern, within any given area, emerges to the eye of the evolutionary biologist, especially a systematist: Related groups of species may display a rather broad range of niche types. And, just as the cichlid fishes apparently show, some rather specialized narrow-niched species may be closely related to one or more very broadly niched, ecologically generalized species all living in the same general area. The generalists occupy a broader range of habitats within the same region that supports a number of specialized kin, restricted to various subdivisions of habitat.

Two antelope groups display this pattern beautifully. They have much to tell us about controls of rates of speciation and extinction: why there are sometimes many, and sometimes relatively few, species in any particular group of organisms. Biologist and paleontologist Elisabeth Vrba has devoted considerable effort to the African antelope fossil record, and to correlating her results with what is known about the species still living. Crucial to her work was the discovery that impalas are most closely related to a group of seven living species of antelopes—a group that includes wildebeests (gnus, the most highly specialized and anatomically most modified of all known antelope, living or extinct).

Impalas are notoriously broad-niched. They eat a wide variety of vegetation and can go for long periods without a dependable source of water. Gnus, bonteboks, and hartebeests are ecologically much more specialized and restricted to particular habitat types (though gnus will undergo long seasonal migrations to track suitable habitat through the year). In a very revealing study, Vrba and colleague Michael Greenacre found that, although there are many antelope species to be found within the confines of the huge Kruger Park in South Africa, impalas account for fully 72 percent of all antelope individuals. Thus, in ecological terms, judging from sheer biomass, the many specialist species taken together are clearly no more dominant or successful in occupying the parkland than that one single generalist species, the impala. Broad-niched or narrow-niched emerge as simply alternative strategies for coping with environmental exigency.

Checking the fossil record, Vrba saw a pattern that is only hinted at by the living species. The wildebeest lineage apparently split from the impala line only 6 million years or so ago, in Upper Miocene times. In that relatively brief time, Vrba has documented the appearance of approximately thirty species of the wildebeest group, including the seven still alive today. During that same interval of time, there have been a grand total of a mere one, or at most, two, additional species of impala, in addition to the single species in the modern African fauna. Antelopes know members of their own species at least in part through the characteristic and idiosyncratic horn shape that the male, or sometimes, both sexes, have. Unlike deer antlers, antelope horns are bony and make good fossils, and Vrba had a more reliable guide to species identity with her fossils than is usually the case with fossils.

Evidently, Vrba concluded, rates of speciation have been higher in the specialist lineage—much higher, in fact, than in the generalist impala lineage. But she also saw that the ranges of species in the geological record were as a rule much shorter for the wildebeest clan than for impalas. The average life expectancy of a species in the impala line is several million years, whereas species in the wildebeest group seldom last more than a million years and often become extinct after a few hundred thousand years. Specialists seem to evolve more quickly. They form new species and accumulate further anatomical specializations more rapidly than their close eurytope relatives. But they also tend to be more prone to extinction.

Paleontologists have known for more than a century that anatomically rather generalized, or simple, members of a lineage usually appear in a greater range of habitats and last for longer periods of time than their more ornate, or anatomically more specialized, close relatives. The usual explanation is that specialists—creatures who concentrate all their ecological eggs in one narrow little basket—are at greatest risk: If a moth species is entirely dependent on a single plant for food resources, or if that plant is entirely dependent on the moth for pollination, one is in desperate trouble if anything should happen to the other. A moth species with several food plant resources, or at least several potential fall-back resources, is much more likely to weather the storm should conditions suddenly turn unfavorable for the continued existence of plants adapted to a particular habitat.

Specialists might seem more efficient at exploiting a habitat when times are good (though the antelope census data reported by Greenacre and Vrba suggest that it may not be so). But when times turn tough it is the generalists—those who, as Stevens points out, are the very ones already used to a wide range of conditions—who will be far more likely to continue to recognize suitable environmental conditions and to keep on living.

But why should specialists actually produce more new species than generalists? Why are there more species in the wildebeest lineage than in the impala line? Speciation is a matter of fragmenting one reproductive community (that is what a species is) into two. Is there something about the mating systems of specialists that render them more prone to fragmentation than their more generalist kin? Vrba, in fact, thinks so. But (to illustrate the diversity of opinion that exists over virtually every interesting issue in science) I am much more persuaded by the old argument that specialists are much more sensitive to nuances in their environment. Specialists detect subtle differences in habitats from place to place. They see a more fine-grained patchiness to their environments than generalists do. If we assume that all species have, on average, an equal chance of becoming disrupted—of a small population becoming isolated from the rest of the species, with reproductive isolation (inability to interbreed) following as a consequence—we can ask: What are the chances of survival of a fledgling species? New species, unless they remain far-flung from their parental species, will tend to be swamped by their parents unless there is something different about them, something ecologically different, that allows them to occupy a different subdivision of habitat. If they remain ecologically similar to the parental species, they will not be able to compete effectively: The vastly greater numbers of the older established, ancestral species are likely simply to swamp the offspring right out of existence.

I think that is precisely what happens to fledgling species that bud off from a eurytopic ancestral species: Their ancestors occur in a wide range of habitats, leaving little chance for an ecological foothold for any descendants that may bud off from them. Stenotopes (specialists), on the other hand, are utilizing only a portion of potentially available habitat, leaving open the possibility that a fledgling species can gain an ecological foothold in some other, similar, yet distinct, subhabitat. If so, the fledgling specialist has a greater probability of survival than the fledgling ecological generalist, even though it is the generalist species, once established, that has the greater chance for truly long-term survival.

Ecological niches—and their relative breadth—thus seem to be at the dynamic heart of the factors that control both the numbers of species represented in any given habitat (true ecological diversity) and the numbers of species that belong to a given evolutionary group (by controlling rates of both speciation and extinction). Ecology and evolution are deeply entwined, even though they represent essentially different basic kinds of systems. Ecosystems cut across the evolutionary grain, with some organisms from this species, others from that, interacting on a moment-by-moment basis as energy flows through the system. Ecological systems are cross-genealogical. Evolutionary systems, on the other hand, are genealogically pure. All components of a lineage are descended from a single, common ancestral species.

These facts suggest that speciation, and especially extinction, are closely tied to the fate of ecosystems, and ultimately to the stability and change of environmental factors. We need, now, to see how these ideas fit in with mass extinctions. But before we turn to actual case histories of the geological past, there is one other clue from modern ecology that we should bear in mind as we look for the common, underlying causes of episodes of extinction.

In their highly influential *The Theory of Island Biogeography*, published in the late 1960s, ecologist Robert MacArthur and E. O. Wilson (renowned ant systematist, conservationist, and pioneer of sociobiology) documented the tight correlation between numbers of species present and the size of a geographic area. The greater the area, the more species present—a trend that is most easily documented when comparing numbers of species on islands of various sizes. The smaller the island, the fewer the number of species present. There are many exceptions, of course: For example, tropical islands will house more species than higher-latitude islands of roughly the same size. Nonetheless, demonstration of the simple relation between area and number of species present has an important and direct bearing as we come to consider the causes of mass extinctions: Changes in size of habitat area have been implicated in many episodes of mass extinction of the remote past.

We now turn to those very mass extinctions of remote geological times. We seek at first simply to understand what happened, what are the actual patterns left in the fossil record by these prodigious events. Only after we find some threads common to all can we consider what constitutes the underlying causes of mass extinction. We will find that the factors controlling diversity on the earth right now have much to tell us about how ecosystems can collapse on a massive scale, cutting across the entire globe, and affecting virtually all evolutionary lineages in existence.

CHAPTER THREE

Biotic Armageddon: Déjà Vu Over and Over Again

THE GLASS MOUNTAINS of Texas are aptly named. The rocks there are chockful of Permian fossils whose shells have been replaced by silica: glass. The fossils are easily etched from their limestone surroundings by soaking them for a while in hydrochloric acid. After years of study by a small army of paleontologists, we know scads of intricate details of marine life in Texas some 250 million years ago. Complex life had been up and running for some 350 million years by then, and the ecosystems whose fragmentary remains emerge from the acid baths seem as complex as any tropical reef system we have today.

Things had been quiet for a long time in those Paleozoic seas. One litmus of peaceful times is the development, as the years roll on by the millions, of creatures that by any standard seem bizarre when compared with their closest relatives. Giraffes, for example, seem a bit odd when compared with their close antelope relatives. Walking sticks, the insects that so closely mimic the stems and branchlets on which they live, are likewise highly modified organisms and a far cry from your average insect.

Highly modified, evolutionary derived creatures do not, of course, appear overnight. The adaptations of these organisms are honed by natural selection and require characteristically rather long periods to take shape. This must be so if one subscribes to the time-honored Darwinian view that selection acts through the ages constantly to modify organismic features, degree by infinitesimally small degree, producing a long trail of gradual modification. But it must be equally true if (as I prefer to see it) complex walking-stick-like adaptations accumulate through a series of species evolved through a string of speciation events. The walking stick is but the last (or, simply, the latest) in a series of species each more modified than the one preceding it. The organisms of each intermediate species, though, would have their own particular adaptations, as well suited to the exigencies of existence as is the walking stick's today. After all, okapis—the rare species *Okapia johnstoni,* still alive, though in small numbers, in the forests of western Africa—effectively represents a stage somewhere between "normal" antelope and the full-blown giraffe.

Those ancient Permian seaways—those that turned out to be the very last of the normal marine environments in what we have retrospectively come to call the Paleozoic—were full of the marine invertebrate equivalents of walking sticks and giraffes. Brachiopods, the dominant shellfish of the Paleozoic, were still flourishing then, so prodigiously, in fact, that the monographs of Smithsonian paleontologists G. Arthur Cooper and Richard E. Grant, devoted to the analysis and description of western Texas fossils, ran to five thick volumes. Brachiopods are still alive today, though in greatly diminished numbers. In Paleozoic times, they were by far the most diverse and numerically abundant form of shellfish.

Like clams, brachiopods have two shells that can be closed water-tight, or left slightly agape to take in waters fresh with oxygen and nutrients. Most species lived rooted to the sea bottom, though some were able to attach themselves to ancient flotsam, floating around the seascape near the surface. Today, brachiopods are found in the interstices of near-shore rocks and reefs, or down in the deep, dark, and very cold waters of the abyssal plains. In the Paleozoic, though, they show up in nearly every conceivable habitat that supported any form of marine life at all, usually in great abundance and species diversity.

Despite their great diversity in shell form, most brachiopods are clearly recognizable as such simply because, rather like clams (or bivalves), their shells consist of two valves. In fact, the trick is usually telling brachiopods apart from clams in the first place. (In brachiopods, one valve is almost invariably smaller than the other, whereas the valves of a clam shell are nearly invariably equal in size and are the mirror image of one another.)

By the time the Permian seas in western Texas were teeming with tropical reef-life, some brachiopods had become rather strangely modified. In one group, the bottom valve became flat and elongated, rather like oysters (a group of bivalves, incidentally, with one valve distinctly larger than the other). The upper valve in these oldhaminid brachiopods was modified into a leaflike structure that could not possibly have completely covered all the soft, internal parts of the animal. Brachiopods are not mollusks at all (as are true clams); they feed and respire with a more or less elaborate gill-like structure (the lophophore) that, in these superficially oysterlike Paleozoic species, apparently secreted most of the strange upper shell as a form of protection and support.

Then there were the richthofeniids, as long and conelike as the oldhaminids were flat and oysterlike. The richthofeniids looked just like the rugose corals that contributed so heavily to the actual structural composition of those Permian reefs. Unlike corals, though, there was an upper lid to close the opening to the deep innards—the lid being the upper shell, rendered tiny in comparison to the hypertrophication of the larger valve. These were indeed the walking sticks of the Paleozoic brachiopod world.

Of course, the oldhaminids and richthofeniids are no longer with us. Neither are any of the species so far etched from those ancient limestones. Something happened to the tranquil seas of the Upper Paleozoic—seas that had remained sufficiently undisturbed long enough for very complex ecosystems staffed by a panoply of organisms, some highly modified, as the two brachiopod groups singled out—to develop and persist for a long time. Anyone who has flown over western Texas and seen El Capitan jutting up some 1,000 feet above the arid floor of the Delaware basin knows how impressive a structure it is. El Capitan is the remains of a single huge Permian reef system, one that must have been around for at least several million years.

The extinction that claimed all these wonderful beasts figures to have been the most pervasive ever (yet, at least) to have hit the globe. David M. Raup—a University of Chicago paleobiologist, who, along with his colleagues Jack Sepkoski and David Jablonski, has contributed mightily to our grasp of the enormity, significance, and underlying causes of mass extinctions—has estimated that possibly as many as 96 percent of all species alive at the end of the Permian perished in this particular extinction event. And, as we saw briefly in Chapter 1, it is these very extinction events, global in scale and generally quite catholic in terms of the diversity of living things they affect, that have been used as empirical datum points in the very act of dividing up geological time. Paleozoic means "ancient life." And Mesozoic means "middle life." This can only mean that life underwent a major change at that point in time when the metaphorical cosmic equivalent of the Times Square Ball (now an apple) fell, ringing out the Paleozoic, all the while ringing in the Mesozoic.

What happened? I have dwelled a bit on some specialized brachiopods, and we have already seen how extinction can indeed claim those species who are narrow-niched and perhaps too specialized for their own good when environmental times change. I have also alluded to the cosmic equivalent of the Times Square Ball, and anyone who has seen a news magazine in the 1980s is aware of the ideas linking mass extinction with extraterrestrial causes, specifically collisions between the earth and comets or asteroids. How do these notions fit together, how do they fit the data, and how do they fit in with the minor host of other ideas clever people have broached as they have grappled with these truly mind-boggling revolutions in the history of life? And what, finally, have they to tell us of our current predicament?

Mass Extinctions: What Are the Facts?

Before plunging headlong into the melee of interpretive theory, we need to establish some basic facts about these mass extinctions of the geologic past. And to do that, we need at least a rudimentary chart of geologic time (see pages xii-xiii)— one that, happily connected so intimately with extinction events themselves, will help locate both us and the major events in the history of life in geological time.

Earth is, roughly, 4.55 billion years old. We know this not from the direct measurement of radioactive decay products in earthly materials but rather from several more roundabout sources: Meteorites, for example, can be radiometrically dated, and their ages tend to cluster around 4.5 billion years. Radiometric dating is based on the known rates of decay of one form (isotope) of an element to another, by emitting one of several possible types of atomic particles as radiation. The amount of the parent isotope in relation to the amount accumulated as the daughter isotope tells a geochemist how long the decay process has been going on; generally, only igneous rocks, cooled from a molten mass, start with a full complement of parent isotope and none of the daughter at the time of the formation of the rock, so, generally speaking, radiometric dating is confined to igneous rocks.

Most meteorites are known to be fragments from the asteroid belt between the earth and Mars. They are assumed to have been formed at the same time that the earth and the rest of the inner planets of the solar system themselves came into being. That being so, it was predicted, and duly found to be so, that the oldest moon rocks would likewise come in at around 4.55 billion years. When predictions are borne out so gratifyingly, confidence in the basic idea—that the earth must be around 4.55 billion years old—naturally grows.

But why are there no rocks sticking out of the earth somewhere that also show that same hoary age? Actually, the age of the oldest rocks yet discovered does keep creeping up on its predicted upper limit: The very oldest rocks now known are from the Canadian Shield, and come in at a very respectable 4 billion years. Canada has recently displaced northern Australia, Greenland, and south-central Africa in the oldest-dates sweepstakes. It is doubtless that some old cratonal (continental) shield area, lying in the nucleus of these or other continental nuclei, will someday produce an even older date.

But in keeping with the Scottish physician-farmer James Hutton's sagacious prophecy, we should not expect ever to find a piece of the earth's crust fully 4.55 billion years old. Hutton, putting the fledgling science of geology on firm observational and deductive footing in the late 1700s, wrote that he saw "no vestige of a beginning, no prospect of an end" to the earth. The earth's crust is a dynamic machine, with rocks beginning to erode the moment they are formed and exposed to the atmospheric elements. Nothing lasts forever, and prospects of finding remnants of the very earliest rocks seem as remote to us now as they did to Hutton.

Though incredibly old hunks of rock have managed to survive stuck in the middle of ancient continental nuclei, we now have reason to believe that Hutton was more correct than he could have imagined: Recycling of the crust through the relatively protracted agencies of mountain building and subsequent erosion is nothing compared to the full-scale and quite literal recycling that goes on beneath the oceans' waves. If we assume, as nearly all geologists do, that the earth has been about the same size throughout its history, then there must always have been ocean basins. But it is only in the past thirty years that accurate dates for the age of the ocean's crust have become available, and they tell an astonishing story.

The job of dating is now, in the main, complete—and the oldest oceanic rocks yet found are a trifling 160 million years old! And these oldest rocks lie outside the deep oceanic trenches now known to be sites where old crust is dragged down and melted, mixing with the deeper mantle, usually never to be seen again back up in the crust. Meanwhile, new crust is forever spewing forth elsewhere, notably at the mid-ocean ridges. Not only our atmosphere but the very bowels of the earth are active with forces constantly reshaping the earthly physical system. Only effectively and long-time "dead" physical systems, such as the moon, are likely to preserve traces of the events in their very formation.

So the earth is 4.55 billion years old, give or take a few million. The oldest sedimentary rocks (rocks formed from mineral grains eroded from other rocks and deposited, usually in bodies of water) are about 3.5 billion years old. Sediments—muds, sands, and the like—are the natural tombs of dead organisms. It is to sedimentary rocks that we turn when looking for fossils. And, sure enough, some of the earliest sediments yet found, dated from associated lavas at about 3.5 billion years, do in fact contain fossils. The fossils are simple microscopic rods and spheres, and look like some of our modern bacteria. Life is indeed a very ancient component of the earth.

But if life was there from a very early stage of earth history, it made no spectacular leaps and bounds toward assuming its present array of splendid diversity until much later. The only fossils other than simple bacteria that have been recovered from the truly ancient rocks of the Archean and Proterozoic have been cabbage-shaped mounds, some several meters high. These are the only traces of life that can be seen with the naked eye for nearly the first 4 billion years of earth history. These stromatolites are actually no more than piles of thin laminations of mud patiently built up, day-by-day, as the months, years, and millenia rolled by. They were formed (as they still are today in a few places) by the growth of blue-green algae (really bacteria) that form mats as the cells divide. All this activity occurs during the day; at night, lying quiescently to await the morning sun, grains of mud and sand accumulate, adding one more layer to the pile. Come dawn, the cells are at it again, growing up and over the last night's muds, forming yet another sticky mat.

For the first 3 billion-odd years of life's known history, that was about it: microscopic bacteria and algal mounds. Yet there was one event, still in the realm of microscopic life, that did occur—not more recently than 1.3 billion years ago. That is the age of the rocks that have produced what Precambrian paleobiologist James W. Schopf and colleagues conclude are the earliest known cells with nuclei.

Nucleated cells, where the deoxyribonucleic acid (DNA) is segregated from the rest of the cell's internal machinery, are found in a host of microscopic, single-celled animals: amoebas, ciliates, and flagellated protozoans, to name but a few of the more familiar groups. But they are also the very same sort of cell found in all multicellular organisms: fungi, plants, and animals. These so-called eukaryotes (true cells, in contrast to the simpler prokaryotic structure of bacteria) have developed a complex division of internal labor: The DNA, which oversees construction and operation of the remainder of the cell but also duplicates itself in the very act of reproduction, is separate from the structures that perform the various physiological processes essential to life, such as assembly of proteins and the processing of energy.

Each of these vital cellular functions is carried on in a distinct and separate structure embedded in the cytoplasm outside the nucleus: Energy production takes place in the mitochondria (in animals; plants have similar structures, chloroplasts, where photosynthesis takes place). Proteins are assembled on completely different structures, the ribosomes, arrayed along the ribbonlike endoplasmic reticulum. Eukaryotic cells are in many ways more efficient, and capable of a greater range of functions, than the simpler prokaryotic cells.

Yet note that prokaryotes—bacteria—are still very much with us. And they are here not just as disease-causing parasites but as free-living creatures in the atmosphere, soils, and waters of the earth's surface. Bacteria can live in places no eukaryote has ever managed to penetrate (so far as anyone knows, at least): As we have already noted, sulphur-fixing bacteria lie at the base of the food chain down in the totally dark hot vents on the oceanic abyssal plains.

Right away we find a theme worth noting in our quest to understand extinctions: Evolution does not bring with it a built-in extinction mechanism, eliminating the old at the advent of the new. When something new comes along in evolutionary history, as when single-celled eukaryotes first appear, they do not by any means automatically supplant those organisms that retain the ancestral features. Even if evolution often represents a new, and, in some instances, even an arguably better (i.e., more efficient) way of doing things, that still is by no means a guarantee that the older, more primitive forms will inevitably drop out, becoming extinct in the face of overwhelming odds in general competition with the new, improved model.

Further evolution, we will see time after time, often seems spurred on by the prior elimination of what came before: Extinction often has provided the very circumstances, some of the very preconditions necessary for further evolution to occur. But the reverse is simply not the case: Newly emerged descendants do not, as the overwhelming rule, routinely cause the extinction (through direct competition) of their evolutionary forebears. If anything it is quite the reverse: As we saw in chapter 2, it is well-entrenched species that in all likelihood stamp out fledgling new species that are not ecologically differentiated enough from their parental species to gain a foothold. If, as dinosaur paleontologist Bob Bakker argues, it does happen that descendant species occasionally outcompete parental species—unto the point of driving them to extinction—it is a minor process contributing to background rather than mass extinction.

Extinction Patterns

We are poised, on our survey of geologic time, at the onset of complex, multicellular life some 670 million years ago. We have not yet encountered the traces of a true mass extinction, though they may well have occurred through life's first 3 billion years. The rise of oxygen content, which some 2 billion years ago changed the earth's atmosphere from a reducing to an oxidizing condition, undoubtedly acted as a poison to the vast majority of the bacteria adapted to the earth's primordial atmospheric conditions. But the record is too hazy for us to discern extinction events in the microbial world over a billion years ago.

Extinction is the ultimate fate of all species. No bacterial clone from Precambrian times is with us today, just as no species of Devonian trilobite, Jurassic sauropod dinosaur, or for that matter Miocene ape is with us either. We speak of the extinction of the dinosaurs, meaning the end of the line for an array of species belonging to various groups within the Dinosauria, at the end of the Cretaceous Period. Yet, by the time they had disappeared forever as the (now) most famous components of the Cretaceous mass extinction, most dinosaur species that have ever lived had already become extinct. As we will see, there is a sliding scale all the way from true, worldwide mass extinction, in which organisms from every conceivable habitat are affected, down to ecological upheavals confined to smaller areas that affect smaller numbers of different kinds of organisms.

But there is also the simple phenomenon of what paleobiologist David Jablonski of the University of Chicago has so aptly called background extinction. Most trilobites were already gone by the time the great Permian extinction finished tri-lobites off for good. Most dinosaurs likewise were extinct long before the end of the Cretaceous. If evolution has no built-in extinction mechanism, weeding out older species as new ones appear on an already crowded planet, nonetheless there is an ongoing process of both speciation and extinction going on within all groups of organisms at all times. New species emerge, and species become extinct.

The two processes may be related: Many paleoanthropologists are fond of the idea that modern *Homo sapiens,* our own species, drove Neanderthals to extinction when *sapiens,* after at least 60,000 years of existence in Africa and the Middle East, finally reached central and western Europe. Then again, the appearance of new species may have nothing to do with the disappearance of old species.

Our focus here is mainly on pattern: what really happens in life's history. And the main point, really, is simply that extinctions are by no means invariably concentrated at certain horizons. All extinctions have a cause: Something must cause population sizes to dwindle, raising the chances of species becoming extinct—for example, forest clear-cutting, or some more subtle environmental change that affects only one or at most a few of the species present in an ecosystem. Invasion by a new predator generally threatens only its potential victims, as when the brown tree snake invaded Guam and began gobbling the eggs of ground-nesting native birds.

For whatever reason, background extinction seems to be going on all the time within all groups of organisms. And that ongoing process has struck some evolutionary biologists, most notably paleomammalogist Leigh Van Valen, as virtually constant for each major group of organisms. According to Van Valen, each group of organisms has a characteristic rate of extinction, one that varies a bit, to be sure, through geologic time, but that seems pretty regular when all extinctions are added up through the total span of time that the group has been on earth. Van Valen has had his critics, both of his interpretation and of his very data on which he based his Law of Constant Extinction in the early 1970s.

Yet there is more than a kernel of truth to Van Valen's patterns: There is most definitely a ticking of the extinction clock that indeed may be fairly regular over long spans of time within groups such as, say, trilobites, dinosaurs, and hominids (our own lineage). I stress this here because we are about to review the last 670 million years of the history of life with heavy emphasis on extinction. And the extinctions that I will stress, quite naturally, will be the prodigious ones—the ones that attracted everyone's attention even in the pre-Darwinian days of the late 1700s and early 1800s.

The danger, of course, in such an approach is creating the impression that all extinction occurs only as mass extinction. I share with a growing number of paleontologists (Stephen Jay Gould most prominent among them) the conviction that much of life's evolutionary history has been contingent on prior extinction of earlier biotas. It has been our concern to establish the importance of extinction—mass extinction, usually without regard to the nature or quality of the adaptations held by organisms.

The alternate vision, one firmly rooted in Darwinian theory, holds that evolution is on the whole a gentle, progressive transformation of lineages, a process that entails, to some degree at least, extinction arising from competition and other forms of interaction between other species. Whatever its cause, extinction is expected, under the traditional Darwinian view, to be a regular, ongoing affair—one that matches speciation and may even display (as Van Valen claimed) a rate typical for each group of organisms. Darwinian theory, in other words, expects background extinction (as well as background speciation). Moreover, the Darwinian vision does not deny the possibility or possible importance of occasional mass extinction.

The question boils down to frequency of occurrence: which has dominated—regular background rates of speciation and extinction as life gradually becomes more diverse and complex through time? Or has the history of life more nearly been a series of mass extinctions—cross-genealogical, ecological collapses, albeit of varying intensities, affecting simultaneously organisms from many different lineages—prompting, as well, subsequent abrupt episodes of evolutionary diversification? Is life's history a stately unfolding tree, as Steve Gould puts it, or a layered series of ecological biotas, one following the other, with characters whose identities change, to a greater or lesser degree, at each extinction boundary?

In arguing that the layered succession metaphor has been underplayed, we should never forget that there is still much validity to the original Darwinian picture. It is not either/or but more a question not only of which picture is the more apt but how the two fit together (and they do fit together to give us the best possible picture of life's history). I stress this issue at this point, once again, because the natural focus on mass extinction as we run through life's most recent 670 million years necessarily draws attention away from the ongoing patterns and processes of background extinction, a phenomenon that we must not lose sight of in grappling with extinction in general.

The Vicissitudes of Life: The Last 670 Million Years

There is something a bit odd about the earliest complex animals to show up in the fossil record. For one thing, the first complex animals seem rather different from modern life forms. Some paleontologists have gone so far as to proclaim this early Ediacaran fauna (named for Australia's Ediacara Hills) literally to be an early experiment in evolution, producing lineages of organisms now wholly extinct and unrelated to the creatures who came to dominate life later on. Others, though, insist that the various different forms can readily be assigned to modern groups, particularly the jellyfish, corals, and sea anemones that constitute the group known as coelenterates.

Many of the Ediacaran fossils resemble sea pens, which are

hydrozoan coelenterates, and it is not unreasonable that complex animal life might have started with a radiation of some of its more primitive elements. But, admittedly, the fossils are hard to interpret. Some that have been taken for early annelid worms, arthropods, and echinoderms may be no such thing at all. We simply don't know. In Upper Precambrian times, similar sorts of fossils show up virtually wherever sedimentary rocks of the right age are preserved. We now have Ediacaran fossils from Namibia, the Charnwood Forest in England, Newfoundland, Siberia, and China, to give a partial list of these truly worldwide occurrences. These creatures persisted for a while, at least 10 million years, and then totally disappeared from the rock record. What happened to them?

Those who see the Ediacaran organisms as members of early groups wholly independent from the origin of the lineages that radiated in the following Cambrian seas (and whose descendants populate our modern oceans) quite naturally see the disappearance of the Ediacaran fauna as history's first mass extinction. After all, the creatures are extinct, and dropped out, it seems, fairly suddenly. Others—myself included—are not so sure. As mobile complex animals (some of which were undoubtedly carnivorous) appeared, both the ecological setting and the conditions for preservation of soft-bodied creatures radically changed. First to come are traces, marks in the sedimentary rock recording the passage of various forms of active, mobile marine invertebrate life. The traces become more complex, until, finally, the first fossil remains of animals with true hard parts—shells and carapaces that readily fossilize—appear.

The origin of mineralized skeletons (independently in a number of major lineages) is an evolutionary and ecological event of grand significance in the history of life. It does, indeed, seem to follow the disappearance of the Ediacaran soft-bodied creatures. But we are interested here primarily in disappearance, and it is the trace fossils that ring the alarm bell when the Ediacaran disappearance is simply taken to be nature's first mass extinction of multicellular life. Organisms plowing through mud and sand disrupt the accumulation of nice even layers. When such bioturbation is relatively slight, a few burrows and trails are found; when it is intense, all traces of bedding disappear too.

Bioturbated sediments are not a good place to look for body fossils generally, and especially not for soft-bodied animals. The suspicion, in other words, is that conditions became progressively less sanguine for faithful preservation of soft-bodied animals. And this particular interpretation tends to appeal to those who suspect that most, if not all, Ediacaran invertebrates indeed do belong to groups with modern representatives: We need not assume that the Ediacaran fauna was wiped out in an early episode of mass extinction.

Yet, on the other hand, according to paleontologist Martin Brasier, in the regions where the sequence of faunas is best displayed (such as in Canada and the Soviet Union), it appears as if the various protoarthropods and annelid worms, together with the sea-pen forms, disappeared before the appearance of the earliest skeletal and well-diversified trace fossils. This circumstance implies that the extinction may well be real, that is, not an artifact of preservation. It also implies that it was not caused by the advent of different forms of complex animal life. It may well be just the opposite: Perhaps shelled creatures arose only after their soft-bodied progenitors had succumbed to extinction. But probably not: That a number of lineages appear to have begun secreting hard shells (by depositing inorganic salts that toughened up their outer skins) more likely reflects a change in the geochemistry of the seawater than a simple occupation of a vacuum left by the older Ediacaran animals. The best bet is that oxygen levels rose high enough to support larger life forms, and that skeletal support systems were necessary for a variety of physiological reasons.

The difficulty in telling what exactly was going on in the Upper Precambrian reflects the remoteness of time, as well as the problems inherent simply in understanding the natures and biological affinities of the soft-bodied forms. Things become a bit easier, at least in terms of judging whether or not there indeed was a mass extinction in the first place, once we pass the Precambrian–Cambrian boundary roughly 570 million years ago. From the Cambrian on up, the organisms are well preserved, and, if now extinct, belong to larger-scale taxonomic groups that still populate the present-day seas. At last, the pattern of extinction and proliferation begins to assume a more regularized and easily detected shape.

Life and Death in the Early Paleozoic

The periods of geologic time usually take their names from places: *Cambria* is the Latin name for what is now Wales. The Cambrian Period is that chunk of geologic time (running from about 570–510 million years ago) that elapsed while the sediments (plus igneous rocks) that comprise the Cambrian System accumulated. The Cambrian System was named by Adam Sedgwick (Darwin's chief scientific mentor, an ordained clergyman in addition to professor of geology at Cambridge University) based on his studies of rocks in Wales.

All Cambrian fossils found so far are of marine creatures: Life appears to have established a serious foothold on land only in the Upper Silurian and Lower Devonian Periods, some 155 million years after the Cambrian began. And as soon as we begin to look over what sorts of marine life populated the Cambrian seas, a minor version of the same debate that marks interpretation of the Ediacaran fauna immediately appears.

There is no doubt that not only all species but a number of larger lineages (higher taxa) that populated the Cambrian seas are now extinct. Trilobites, the earliest and most primitive of true arthropods (the phylum of joint-legged creatures, including, among others, crustaceans [crabs, lobsters, shrimp, and kin], as well as the great insect and spider groups) dominated most known Cambrian ecosystems. Trilobites are fairly common fossils in Paleozoic rocks. With a well-developed head equipped with eyes, a middle segmented trunk region, followed by a solid tail piece, trilobites were active, mobile members of the bottom-dwelling ecological communities of nearly all Paleozoic marine environments.

Although trilobites were to persist until the very end of the Paleozoic, for some 265 million years after the Cambrian ended, they were never again to be seen in the profusion they developed in the Cambrian. Counterposed, however, to the great species diversity of trilobites in the Cambrian was the relative lack of what Stephen Jay Gould (in his *Wonderful Life*, centering on the Cambrian fauna of the Burgess shale) has called disparity: There was, at least to my eyes, a greater array of purely anatomical diversification in the post-Cambrian trilobites than ever showed up during the Cambrian. This implies a greater diversity of ecological roles in the later trilobites. But, admittedly, it is really difficult to tell what roles trilobites in general played in the local ecosystems of their times.

Except for the unusual circumstances of the very few units like the Burgess shale, where appendages of trilobites and a plethora of other arthropods are faithfully preserved, trilobite appendages are simply never found. Like most soft-bodied creatures, trilobite limbs were simply too prone to decay. Lacking the heavy mineralization typical of the upper side of the trilobite body (including the eyes), the legs and antennae were almost always destroyed. And that is a disaster from the point of view of specifying trilobite economic roles: Arthropods swim, walk, burrow, feed, respire, defend themselves, and reproduce with their appendages. Trilobites were no exception. We simply do not know how Cambrian ecosystems could support so many different species of trilobites, many of which looked quite a bit like one another, and all of which (lacking any detailed evidence) are assumed to have been scavengers and detritus feeders, meaning they scurried about the seafloor, roiling up the bottom and consuming small bits of organic debris and, in all likelihood, living microorganisms.

John J. ("Jack") Sepkoski, who has compiled a tremendous

paleontological data base that has contributed so much to the actual documentation of patterns of extinction (and has served, as well, as a point of departure for a consideration of the causes of extinction, as we shall see), has written of the origins and fates of what he calls evolutionary faunas. Sepkoski sees the marine invertebrate faunas of the past 570 million years breaking down into three distinct entities: the Cambrian evolutionary fauna, the Paleozoic evolutionary fauna, and the Mesozoic–Cenozoic evolutionary fauna. These were the three main clusters, or associations, of families of marine animals that came out of an initial computer analysis of the distribution of marine animal families over the past 570 million years.

These faunas are neither ecologically nor temporally separate from one another. Sepkoski sees elements of the Paleozoic fauna already present in the Cambrian and vice versa: Elements of the Cambrian fauna persist till the end of the Paleozoic. It's just that the typical composition of Cambrian communities is so unlike those of the next succeeding Ordovician (and later) Periods. With trilobites were a few primitive mollusks, inarticulate brachiopods (meaning that their hinge lines lack teeth), a few articulate brachiopods (those with teeth and sockets on their hinges—the vast majority of brachiopods, and the early progenitors of the bizarre forms from the Upper Permian), sponges, and members of a few other taxa, long since extinct, whose biological affinities are still open to question. For example, archaeocyathids, which were virtually restricted to Cambrian seas, were once considered to be an extinct form of sponge (which they vaguely resemble in their conical shapes), but are now thought to be calcareous algae.

Small wonder, then, that some paleontologists, following the same line of thought favored by some with the Ediacaran fauna, have viewed the Cambrian fauna as an early evolutionary experiment—one that, erased by extinction at the end of the Cambrian, was replaced by a fauna much more nearly like the marine life we have today. Yet Sepkoski's Cambrian fauna, remember, is a statistical association of different major groups of animals, elements of which persisted long after the Cambrian.

What sets off the Cambrian most radically from the later marine bottom communities, at least in my view, was the addition of a major category, an ecological cadre, that was barçly present in Cambrian times. What sets off the Ordovician and the rest of the Paleozoic from the Cambrian is not so much what is lost from the Cambrian as what is added to later faunas: a veritable explosion of marine invertebrates, all of which sat on, or were directly attached to, the sea bottom. These were all creatures whose skeletons were formed of sturdy calcium carbonate. They were content to sit there quietly on the seafloor, filtering nutrients from the surrounding water. Their advent, in the early Middle Ordovician, is coincident with the spread, for the first time, of major seaways over the continental interiors—seaways whose bottom sediments were predominantly of limy particles.

Nearly all the cast of characters of this vast array of bottom-living, calcareous-shell-secreting, filter-feeding invertebrates had known progenitors in Cambrian seas. The major characters in this new ecological play included crinoids, or sea-lilies—stalked and stabilized relatives of sea urchins and starfish. Crinoids today are restricted to the nooks and crannies around reef systems and to the remote depths of the dark abyssal plains of the oceans. In the Paleozoic, though, they were abundant in most shallow-water marine ecosystems. Bizarre for animals, crinoids (and their close relatives, including blastoids) are generally rooted to the seafloor by a long stalk composed of a series of many tiny plates. The main body of the crinoid animal lies in a cup on top of the stalk. Long, feathery arms extend upward from the body, filtering food particles from passing currents of seawater.

Bryozoans were another important group to appear and play an important role in the new ecological theater. Bryozoans are often called moss animals because they form colonies composed of hundreds, even thousands, of tiny individuals that may cover a wharf piling or blade of algae reminiscent of the way moss grows on trees. Despite the tiny size of each individual bryozoan animal and their colonial mode of life, bryozoans are actually close relatives of brachiopods. Like brachiopods, bryozoans have a feathery tentacle system, the lophophore, that is used for feeding.

Articulate brachiopods, descended from their simple orthid lampshell Cambrian ancestors, exploded into an array of different forms, and came, for the first time, truly to dominate life on the seafloor. Several major groups of corals—all of a sudden in abundance in the Lower Middle Ordovician seas, in stark contrast to the several specimens of coral known so far from Cambrian rocks—also rose to prominence. Corals, like bryozoans, brachiopods, and crinoids, derive their living primarily by filtering food particles from seawater. Other animals—most notably, vagrant scavengers and carnivores—also proliferated in the Ordovician, far beyond their modest presence in the Cambrian. Nautiloid cephalopods, molluskan carnivores related to squids and octopi, proliferated a bit earlier than the bottom-dwelling filter feeders; they, too, had only modest Cambrian diversity. Their direct descendant lives on as the several species of the chambered *Nautilus* of the modern Indo-Pacific region. Other groups, such as jawless fishes (which show up in Middle Ordovician times), have more dubious connections with known Cambrian fossil taxa. The lampreys and hagfishes of the modern world are highly modified survivors of these early forms of fish.

What we have, then, in Ordovician times is a great expansion in ecological diversity in benthic communities. Why this sudden explosion of undersea species? We are presented with something of a chicken-and-egg dilemma: The expansion of all these shelled bottom-dwelling invertebrates corresponds precisely with the first truly widespread blankets of carbonate sediments over the continental interiors. Seas had begun to encroach widely over the continental interiors by Upper Cambrian times, but it was in the early Middle Ordovician that the

benthic = occurring at the bottom of a body of H₂O

bottom sediments of these seaways became predominantly limy. But which came first: limy environments or the profusion of limy organisms?

Limestones and limy shales are largely derived from the shells of marine invertebrates. Did the creatures simply create their own substrate? Or was there, in a direct geochemical sense, a greater availability of lime—one that is reflected in both biological diversity and in the widespread occurrence of limy seas? I suspect the latter, however vague the notion remains. Like the initial proliferation of skeletonized creatures, an event that affected many different lineages nearly simultaneously, and that cut across both evolutionary and ecological lines, the sudden burst of diversity of bottom-dwelling, filter-feeding, well-skeletonized marine invertebrates more likely reflects an opportunism occasioned by environmental (presumably chemical) change than it marks a reaction to previous extinction.

Taking the Measure of Cambrian Extinctions

Cambrian times saw many extinction events. The Upper Cambrian, in particular, has become celebrated in recent years for its patterns of abrupt extinctions, followed by waves of new invaders that appear and set up ecological shop. Allison R. Palmer coined the term *biomere* to describe the pattern. Biomeres are bodies of rock housing a distinct evolutionary cycle of their constituent fossilized organisms. New species appear relatively abruptly near the base of the biomere. They proliferate into a number of different lineages. Then one or more waves of extinction wipes them all out, marking the end of the biomere, and the beginning of the next one.

There are many exciting aspects to the biomere patterns. The first is simply the regionalization of extinctions: Though biomeres might well be intercontinental or even worldwide in scope, the original concept came about through detailed studies in the Great Basin of western North America. Here, in Upper Cambrian times, the sea lapped in as far as Minnesota and Wisconsin; traveling westward, shallow-water (with limy bottoms) communities now preserved in mountain ranges in Utah gradually sloped down toward the abyss, now preserved in rocks of the same age to the west in Nevada. The cycle of extinction, followed by invasion, then by evolutionary proliferation, and finally by another episode of extinction, is thought to have affected mostly the trilobites and other marine creatures inhabiting the shallow-water environments. Deeperwater faunas were less affected. Here, deep down in geologic time, way back in the Cambrian, there is clear-cut evidence for selective extinction not so much by taxon (organismic group) but by environment and geographic region.

selective ext. by environ.

Biomeres show us something else, as well: Extinctions can be deceptively difficult to measure. What is obviously an important extinction event to a field paleontologist may not look like much to someone who performs a statistical analysis on a mound of data compiled from many different sources. The field paleontologist can put a hammer right on a spot that marks the final appearance of a goodly number of species. Above that spot—that line, actually—there will be other fossils, some often quite similar to those below the line.

Nor can a field paleontologist assume that local disappearance equals extinction. Anyone who has spent much time collecting fossils knows that it is environmental change, factors that change the geographic distributions of species, that account for the vast majority of comings and goings of fossil species. And local disappearance has vastly different effects, depending on whether an entire species, or just a part of one, becomes extinct. Only when the field paleontologist goes back to the laboratory, the literature, and the comparative museum specimens can it be accepted that the point of disappearance found in the field truly does represent the terminal occurrence of a number of species—in other words, true extinction.

But it is frustratingly difficult to put together a regional—let alone continentwide, intercontinental, or global—data base on extinctions. Species are the fulcral elements of extinction. Of the perhaps hundreds of millions of species of organisms that have existed since the Cambrian, only a small fraction are yet known (or ever will be known) as fossils. Yet the number of fossil species so far described is prodigious. Confounding the problem are the wide differences in taxonomic procedure (and competence) that have marked different times, and countries, in their paleontological pursuits. It is difficult to place equal faith in all paleontological systematic descriptions of species. Thus the data on fossil species are, in a sense, too few, though there is too much to handle. And the quality of those data is uneven (though the difficulties inherent in recognizing true biological species in the fossil record have been greatly exaggerated).

Several years ago, Dave Raup (one of the three preeminent extinction-minded paleontologists at the University of Chicago) looked into the problem of data gathering for the documentation of extinction patterns. Raup clearly saw the problems of capturing the data at the species level. But he also realized that it is at the species level that such extinctions, in fact, take place: Species are literally at the interface between process and history. Raup then looked at the other side of the taxonomic ledger, the truly higher-level taxa: the phyla, classes, orders, and families. Two effects emerge and become stronger the higher you look in the Linnaean hierarchy.

diff. levels of ext.

First, the drama of the effect of the extinction is enhanced: An extinct phylum is more impressive and biologically more significant than the extinction of a species. Phyla are the largest divisions of life after kingdoms. Arthopoda, Mollusca, Chordata (our own) are examples. That extinction of a phylum is more significant than that of a species derives from the generally safe assumption that a phylum is a lineage with many thousands, even millions, of species in it. This is true, for instance, of the Phylum Arthropoda, which contains insects along with a vast array of other major groups. Some phyla have been named, especially from the fossil record, based solely on their presumed remoteness from other known taxa, and have very few species allocated to them.

The other effect of looking at phyla and other higher-level taxa has to do with sampling: Species may be the focal point of extinction and evolution. But you can easily miss a species by never finding it, or even not seeing it in the literature. In contrast, all you need is to find a single one of the thousands of component species to document the existence of a phylum. Phyla are easy to sample; species are difficult.

The problem, of course, is that not only does the extinction process take place at the species level but also very few phyla, or even classes, have ever become extinct (though some paleontologists have claimed otherwise). Measuring extinction at the higher levels is so coarse that the entire phenomenon of extinction will inevitably be underestimated. Because no kingdoms are known ever to have become extinct, measured at the highest levels, extinction would appear never to have taken place!

Raup suggested a commonsensical, golden-mean type of solution: Pick a taxon level in the middle—one that is relatively easy to sample (all you need is one species . . .), but one not so far removed from the species level that the dynamics of the actual extinction process is actually missed. Raup suggested families; Jack Sepkoski's data base, begun independently from Raup's analysis, is at the family level for precisely these pragmatic, split-the-difference sorts of reasons. The data on families in the fossil record are reasonably good, and families aren't so far removed from the extinction process itself—even when they don't hit higher taxa—that important extinction events are less likely to be missed. It is a compromise. Sepkoski is currently building a species-level data base, one that will pick up all levels of extinctions. It is a gargantuan task.

There is one other problem in measuring extinctions that has proven to be rather a bone of contention. The issue involves some fundamental notions of the very nature of evolutionary lineages. During the past twenty-five years or so, systematists have tended (very strongly) to adopt a strictly genealogical approach. This means that every group—say our own Family Hominidae—must consist of only the ancestral species and all (including our own species, *Homo sapiens*) that have descended from it. That sounds perfectly plausible.

The problem arises with older classifications. For example, we were conventionally classified as the only species in the Family Hominidae (until fossil humans began showing up, and they, naturally enough, were also considered to belong to the Family Hominidae). However, there is the Family Pongidae to consider. These are the great apes, in traditional classification, that is. The problem is that some of those great apes are more closely related to us than they are to the other apes: We share a more recent common ancestor with some of those apes than with others. As a kind of ape ourselves, it will hardly do to separate ourselves and throw all others, regardless of kinship with us, in with each other. That violates the principle that all descendants, and only those descendants, of a common ancestor should be classified together.

Apes are now, by near-universal agreement, members of the Family Hominidae. We no longer recognize a Family Pongidae—that is, a group consisting of apes but excluding ourselves. Evolution creates genealogically pure (so-called monophyletic) lineages. It is not mere language but the geometry of the history of life that is the issue here.

These issues, as I have said, have begun to affect the analysis of extinction patterns. What does it mean if a Family becomes extinct, if that family is not actually monophyletic? What if the current crop of great apes had become extinct during the Pleistocene (it could have happened: orangs, pygmy, and true chimps, as well as mountain and lowland gorillas, are currently facing extinction). In our hypothetical example, to record the extinction of the Family Pongidae would have been to record an evolutionary nonevent, because one member of the Family—ourselves—had so far managed to avoid extinction. Remember this is a thought experiment; thankfully, the various species of *Pongo, Pan,* and *Gorilla* (not to mention yetis and so on) have in fact so far avoided extinction. We are only asking: What would be the effects on our data had these species become extinct in the Pleistocene and had we persisted in calling them Family Pongidae? We would clearly be in error, claiming that a Family became extinct when the Family had been erroneously constituted in the first place.

Against this objection to the unreliability of data collected in the days before all taxa were construed to have to be monophyletic in the modern sense is the sensible rejoinder that recording the extinction of Pongidae would actually note the fact that some four or five species indeed had become extinct. If we had stuck to the family level, using evolutionarily correct nomenclature, we would have missed recording the extinction that we alone had survived: The family would still be extant, albeit represented by a sole surviving species, and consequently we would have failed to note the extinction of all the other species. Genealogically incomplete (nonmonophyletic) groups often pick up events at the species level that would be missed if the groups were scrupulously monophyletic! The solution really is to get those species-level data that will also tell us when (monophyletic) lineages (of whatever rank, from genus on up through kingdom) actually also become extinct.

Mass Extinction in the Cambrian

The events at the end of the Cambrian are not in the select inner circle of the truly great mass extinctions. The five biggest extinction events so far in the history of complex organisms occurred later: at the Ordovician–Silurian boundary, near the end of the Devonian Period, at the Permian–Triassic boundary (the Paleozoic–Mesozoic boundary), at the Triassic–Jurassic boundary within the Mesozoic, and at the end of the Cretaceous (the Mesozoic–Cenozoic boundary). Nonetheless, according to Cambrian trilobite paleontologist Stephen Westrop, ten (nearly half) of the trilobite families present in North America became extinct at the end of the Cambrian 510 million years ago—no trivial number from the vantage point of both trilobite evolutionary history and the composition of contemporaneous shallow-water marine ecosystems. The Cambrian events were of an intermediate scale, and some of the details of their pattern of occurrence shed further light on extinction phenomena generally and on the difficulties inherent in their interpretation.

The temporal aspects of extinction events are, of course, crucial to deciding which of the many possible different factors were actually responsible. Because the celestial impact hypothesis has received so much attention in recent years (not least because discussion has centered on the Cretaceous–Tertiary extinction event, which took out the dinosaurs with so many other forms of life), the idea that even the greatest, most sweepingly global mass extinctions were virtually instantaneous has gained wide circulation. We are reportedly currently losing species at the rate of about one a day (some say many more)— appallingly fast, but nothing like a scenario that sees millions of species literally dropping out overnight. One problem we face in locating our own environmental difficulties in the sweep of earth history is matching up what we think we see going on today with our best, most accurate assessments of rates of extinction in the geologic past.

The data for Cambrian trilobites tell an interestingly mixed story. The extinction at the end of the Cambrian was simply the third and last of the extinctions that terminated the three Upper Cambrian biomeres. In their classic development, in the Great Basin of western North America, where Palmer first studied them, the turnover between the top of the old biomere and the base of the new is indeed sharp. In some places, there is a distinct change in the rocks themselves, indicating environmental change; in most others, though, there is no such switch in sediment type as the trilobite fauna jumps from one assemblage to the next.

With no obvious disturbance in sedimentation, there is no reason to assume that there is a gap in the time preserved in the rocks. And it is reasonable to conclude that the event—disappearance and replacement—was very quick indeed. If it was not literally overnight, then at least the turnover was so rapid as to appear instantaneous: a few hundred or even a few thousand years can look instantaneous in the Paleozoic rock record. And even a few hundred years fall within the range of ecological time: The imperiled old stands of redwoods of the Pacific Northwest are themselves hundreds of years old. Ecological time (ranging from perhaps minutes on the low end to centuries, perhaps a millennium or two, at the upper end) is simply the time scale in which ecological processes (predators taking prey, changes in population densities and the geographical distributions of species, etc.) characteristically take place.

But later work has changed the picture somewhat. Westrop and his colleague Rolfe Ludvigsen, for example, have found that the trilobite extinctions occurred sporadically throughout as much as 26 meters of rock in some places. As we saw with the chalk layers of Stevn's Klint, it is never easy to put actual temporal numbers on accumulations of sediment except in rare circumstances, as when thin layers (varves) are known to have accumulated at the rate of one per year. In such situations, it's easy: Like tree rings, all you need to do is count the layers and you have the elapsed time in years.

But most sedimentary rocks have no such internal structural clues to tell us how long it took for the sediments to accumulate. One tremendous wave of muck, crashing down the continental slope in a turbidity current, can build up a thick pile in a matter of minutes (or hours—it takes a while for the finer-grained sediments to settle out). But most sedimentary rocks have some bedding (internal horizontal layering) that testifies to the slow, steady and rather gentle accumulation of sediments over millennia. It's just that the layers are not usually formed daily but rather may take years or even millennia to build up. The red clay found on the bottoms of parts of the world's deep oceanic basins as a rule accumulates at the incredibly slow pace of 0.1 or 0.2 centimeters per year. The rate is so slow because the clay particles are mostly cosmic dust coming in from outer space.

So we don't know how much time it took to accumulate what is now 26 meters of Upper Cambrian rock (which is a compacted version of the original even thicker pile of sediments), but it surely did not happen overnight. Nor do the trilobites occur throughout, and disappear only at the top of the pile. That's the whole point: Trilobites are dropping out throughout those 26 meters. In such places, the extinction seems to have been a more gradual, and certainly a more protracted, affair than the areas in the Great Basin that Palmer originally studied.

[margin note: Does he mean trilo species?]

Here lies a clue in our search for the causes of extinction. Ludvigsen and Westrop think that a simple environmental change—a rise in sea level—made near-shore environments less patchy and more homogeneous. Offshore, rising sea level had little effect. Increasing environmental homogeneity implies loss of niches. Such a change in nature and area of accustomed habitat seems to underlie the patterns of Upper Cambrian extinction. Disruption simply occurred more rapidly in some places than in others. As we turn to the major mass extinctions that, in Upper Cambrian times, still loomed ahead for the world's ecosystems, we will continue to garner hints of the probable causes of mass extinctions.

CHAPTER FOUR

Patterns and Clues in
Paleozoic Mass Extinctions

WE HAVE REACHED the point, a bit into the Ordovician Period, where life has assumed a more or less familiar organization, albeit with a cast of characters that, by modern standards, is decidedly archaic. The corals of the Paleozoic are not the corals of today; the snails and clams had a more primitive leer than their counterparts of today; trilobites are no longer with us. And so on. But, after the Cambrian, it is simply much easier to look at the remains of ancient shelly sea bottom communities and think of them in the same general terms that we routinely bring to bear on all modern communities: herbivores and carnivores, filterers and scavengers, photosynthesizers and decomposers. And before the Paleozoic was to progress very much further, plants, arthropods, and vertebrates were to pioneer life on land, establishing the first subaerial ecosystems. Some of the extinctions that were to come affected life on land more than in the sea, others vice versa, and the very biggest, of course, left no habitable corner of the earth untouched.

I have spoken briefly of the near-rhythmic pattern of monotonous, basically stable systems of ecological communities inhabiting the Paleozoic seafloors for millions of years only to be overturned and replaced in due course by successor ecosystems of varying degrees of resemblance to their predecessors. Resemblance is judged on how closely related (and apparently ecologically similar) the species in the succeeding systems are to those whose organisms staffed the preceding systems. In general, the more geographically and taxonomically pervasive the collapse, the less similar the successor ecological systems were to the ones preceding them.

It is entirely reasonable to consider the evolutionary history of life as governed, to a very large extent, by ecological disruption dislodging the comfortably ensconced: Background speciation and extinction goes on during times of relative tranquility. More massive across-the-board extinctions free up ecological space, nurturing and perhaps even goading new species to appear and survive. These new species are composed of organisms of a novel adaptive stripe that form the component parts of new ecosystems. Indeed, that is the basic picture of life's history that I have painted in my own earlier book *Life Pulse*.

But it is not extinction's importance as a goad to further evolution that concerns us most at this juncture. Rather it is the pace and texture of extinction itself as a prelude to exploring the causes of ecosystem collapse. As we survey the five most devastating extinctions since the advent of the Ordovician biota, we must simply take the cast of characters—the species, and the higher taxa in which they are arrayed—as given. We seek here not to understand where they come from but how they disappeared.

Extinction in the Upper Ordovician

According to paleontologist Patrick J. Brenchley (in articles written in 1989 and 1990), some 22 percent of known marine families became extinct in the Late Ordovician. Statistically, as Brenchley says, this Upper Ordovician event qualifies as a mass extinction. In some groups, apparently more than 50 percent of the included species became extinct. Yet it is obvious to even the most casual observer that, from an ecological point of view, the Ordovician–Silurian boundary was hardly the point of major ecological revolution that some of the later extinctions—Permo–Triassic, Cretaceous–Tertiary, especially—would prove to be. Though 22 percent of Upper Ordovician families did, rather abruptly, disappear, larger-scale taxa were not affected, and the repopulating of ecosystems in the early Silurian was achieved with organisms that bear a direct resemblance to their Ordovician progenitors.

Yet the Ordovician extinction was truly a worldwide phenomenon, and one that affected most if not all of the varied groups that had come to populate the Ordovician sea bottoms. Brenchley paints a picture of moderately rapid diversification of the sedentary, filter-feeding organisms that lie at the heart of marine invertebrate communities ever since the early Ordovician. By Mid-Ordovician times, diversity had reached a peak. As the Ordovician Period wore on, many of the groups show a progressive decrease from their Mid-Ordovician diversity highs: for example, Brenchley tabulates more than 200 genera of Mid-Ordovician brachiopods, a number that decreased to about 150 in later Ordovician faunas. There are still fewer genera in the Lower Silurian. The same sort of progressive decline is seen in the carnivorous nautiloids (cephalopod relatives of squid and octopi, still represented today by several species of pearly nautilus of the western Pacific). Trilobites, too, declined from more than 200 genera in the Middle Ordovician to fewer than 100 in the Lower Silurian. Crinoid and other echinoderm families also became fewer in number toward the end of the Ordovician.

The Upper Ordovician was also a tough time for two groups of organisms with important fossil records: the graptolites and conodonts. Both groups lived life up in the water column, well away from the seafloor. Graptolites, now totally extinct, were colonial floaters, very possibly closely related to our own chordate lineage. They were colonial creatures, with each individual animal living in a separate compartment arrayed along a sawtoothed branch; each colony consisted of one or more branches. The very term *graptolite* means "written in stone": graptolite fossils usually look like carbonized pencil sketches on the dark, flat surfaces of Paleozoic shale.

Conodonts—minute, toothlike fossils—were parts of creatures whose otherwise wholly soft anatomy is only rarely preserved. Their affinities remain obscure, though recent work points to chordate affiliations as well, meaning that they, too, could well have been distant relatives of our own line of descent. Both conodonts and graptolites have figured heavily in paleontological research because they have proven so useful in correlation: telling two far-flung bodies of rock are at least roughly of the same age because the fossils they contain are either identical or at least closely similar. Because these creatures swam (conodonts) or floated (graptolites), they got around; they also became extinct and evolved at decently brief intervals: Relatively short life spans (i.e., of species or genera) and broad ranges are the ideal requisites for using fossils in rock correlation studies.

I dwell on graptolites and conodonts a bit because their data are at the species level—at a critical level, in other words, where whatever is causing extinction is actually working. We can measure the Upper Ordovician decline of both conodonts and graptolites in terms of actual numbers of species—the crucial level for seeing what is really going on. The echinoderm data are in families, the by-now traditionally accepted midrange level, where practicality intersects with expectations of precision and a desire to sample at the smallest taxonomic levels possible. On the other hand, the trilobite and brachiopod data are for genera—the data are really beginning to improve!

Some groups do not show the pattern of gradual decline in diversity from the Mid-Ordovician through the Lower Silurian that is seen in the brachiopod data. Bryozoans—the colonial brachiopod relatives that have been such important ecological members of the marine biota from the early Middle Ordovician right up to present times—show no such decline. Nor do family- and higher-level data show much in terms of a global extinction event for bryozoans at the end of the Ordovician. Clams, too, fail to develop the pattern to any marked degree.

Other groups, such as the two branches of Paleozoic coral-dom (tabulates and rugose corals), show an increase in generic diversity from the Middle Ordovician to their collective peak in the Middle Devonian. However, in the late Ordovician, corals underwent a major pruning: Brenchley reports that about 75 percent of tabulate and rugose genera in the Upper Ordovician never made it into the Lower Silurian. And so on: Even if a group does not show a progressive decline in diversity from the Middle Ordovician through the Lower Silurian, it still may well show a dramatic cutback toward the end of the Ordovician. This is so even for the bryozoans, as some regions, at least locally, show drastic extinction. It may well be that bryozoan extinctions were as global as they were for other groups, and we just don't know it because the data are not yet sufficiently refined.

So far, the picture looks like a regular decline in diversity, with little to suggest even a relatively abrupt, major extinction event at the end of the Ordovician. And there is little doubt that progressive diversity decay is an essential ingredient in the Ordovician–Silurian boundary extinction. It has been notorious for well over a century now that Silurian faunas are well-nigh cosmopolitan. The same species is apt to show up on two continents. The spectacular Middle Silurian trilobite *Trimerus delphinocephalus* from the neighborhood of Rochester, New York, is a dead ringer for *Trimerus delphinocephalus* from western England, so much so that early paleontologists, much as they liked giving different names to fossils only slightly separated in space, time, and anatomy, felt compelled to give their specimens the same name.

Regional endemism declined because various pieces of what were later to emerge as Europe and North America were beginning to get pretty close to one another. Britain and other parts of Europe began to approach what is now Scandinavia, and the proto-Atlantic Ocean (sometimes called Iapetus) was shrinking as Avalonia (whose remnants now form eastern Newfoundland) approached the core of what is now the North American continent (Laurentia). As these continental bits drifted ever closer to one another, old barriers to marine life began to crumble and the regionalization of faunas so prominent in Middle Ordovician times progressively declined.

But was there an actual, single terminal Ordovician extinction event? According to Brenchley, the answer is yes and no. No, there was no single, instantaneous and worldwide simultaneous collapse of marine ecosystems. But yes, in the Late Ordovician, extinction was especially intense and worldwide in scope. And extinction occurred in two distinct steps, separated by 1 or possibly 2 million years. The distinction between the two phases of the extinction is best documented in Europe; the limestones of North America have yielded a more composite picture, where the two phases are difficult to distinguish.

Of the two waves of extinction, the European data clearly show that the first was the more severe. The last million years or so of the Ordovician saw a widespread fauna dominated by brachiopods (the so-called Hirnantian fauna). Nowhere near as rich in diversity as the pre-Hirnantian Upper Ordovician faunas, this terminal Upper Ordovician fauna hung on over much of the globe until the second wave completed the job of the first, and Ordovician life was over. Paleontologists have long suspected that the Hirnantian was a very cold-water fauna: our first clear signal of a specific underlying extinction cause, and a theme that is soon to reappear.

Anyone who has ever collected the incredibly abundant and beautifully preserved fossils literally falling out of the rocks along the roadsides of western Ohio and eastern Indiana knows just how extraordinarily diverse the Upper Ordovician marine communities actually were. No one knows, yet, whether the uppermost Ordovician of this midwestern region is pre- or post-Hirnantian—that is, whether the disappearance of the Upper Ordovician fauna in the American Midwest reflects the first or second wave of extinctions as documented in Europe. But go to Dayton, Ohio, and look at the Lower Silurian after looking at the Upper Ordovician around, say, Madison, Indiana, and you will have a compellingly graphic image of the change a mass extinction can wreak on ecosystem complexity.

Extinction in the Upper Devonian

Waldron, Indiana, is a modest farming community some 40 miles southeast of Indianapolis. At first glance, there is little to distinguish Waldron from all the other little towns of the fertile Midwest except the cement quarry just south of town. Waldron long ago lent its name to a body of limy mud—the Waldron shale—that shows how spectacularly marine life had rebounded after the extinctions at the end of the Ordovician. The fossils of the Waldron are just about as prolific and diverse (and as beautifully preserved) as those last seen as the Ordovician was drawing to a close.

Ordovician life was dealt a blow that was far from fully mortal. In some places, it took a good 5 or even 10 million years for the seafloors to become as densely littered with the shells of brachiopods, clams, and snails, the skeletons of bryozoans and trilobites, and the calyces of corals as they had been prior to the Upper Ordovician extinctions. But those sea bottom communities eventually did come back. And though some groups were no longer around, the faunas of the Silurian and succeeding Devonian really do have the same basic structure and feel that had first appeared back in the Lower Middle Ordovician. The groups that were eliminated—some Ordovician trilobites, for example—seem not to have been replaced by newly evolved groups. There was a reproliferation based on what survived, and no major new groups, or forms of adaptive organization, evolved in seeming response to the Ordovician cutbacks.

The Silurian segues smoothly into the Devonian, so smoothly that it has proven impossible so far to reach international agreement on where the Silurian ends and the Devonian begins. Faunas became a bit more isolated and parochial in the Devonian from their heights of cosmopolitanism reached in the Silurian. But it wasn't until the Devonian had gone on for some 40 million years that the next major extinction wave hit, vastly disrupting life once again on the seafloor and, for the first time, on land as well.

The first thing that strikes one as odd about the Upper Devonian mass extinction is that, unlike all the other major, worldwide events, it did not come at the very end of a geological period but rather between the last two Devonian epochs: the Frasnian and Famennian. Named for towns in Belgium where rocks of Upper Devonian age are particularly well exposed and studied, the Frasnian and Famennian are starkly different in terms of the ancient faunas they contain. The Frasnian was the last great heyday of exuberantly rich Devonian marine faunas whereas the Famennian records a severely restricted set of ecosystems, with far fewer numbers of species dominated by brachiopods, clams, and glass sponges.

A second glance, though, shows the pattern of Upper Devonian extinctions to be closely similar to those we have just seen at the end of the Ordovician: the Hirnantian, that terminal Ordovician fauna, too, had few species compared with the preceding Ordovician faunas. The major extinction was over by the end of the Frasnian. What was left was a strange vestige of Devonian life rather than a clear-cut redevelopment of normal marine ecosystems. When normal marine communities did eventually reappear, their fossils were traditionally taken as the base of the Upper Paleozoic Carboniferous Period.

Brachiopod specialist George McGhee, Jr., has been instrumental in bringing the events of the Upper Devonian in as clear a focus as has proven possible so far. He has shown that throughout the latter half of the Frasnian, some 3 or 4 million years, extinction rates were up. There was no single point in time where tremendous numbers of species synchronously disappeared, though there seems to have been a pulse of extinction at the very end of the Frasnian.

McGhee has recorded a number of fascinating ecological patterns in conjunction with events in the Upper Devonian. All signs point to a rather marked drop in worldwide temperature. In what McGhee calls the latitudinal effect, for example, he notes that tropical marine ecosystems, including reefs, were especially hard-hit. Massive reefs composed of tabulate corals and stromatoporoids (traditionally considered forms of coral, but now thought to be sponge relatives) had long been a staple of tropical Silurian and Devonian ecosystems; they disappeared altogether in the Frasnian.

Among brachiopods, some 86 percent of all Frasnian genera failed to make it into Famennian times. But of those that did get across, McGhee notes that the vast majority were cold-water-adapted forms. Once again, there seems to have been a differential extinction bias against warm-water-loving marine creatures.

There was also a bias of land versus marine survival: In an argument that mirrors George Steven's ecological ideas (chapter 2), McGhee points out that far fewer lake-dwelling fishes became extinct than was the case among their close relatives in the marine realm. He attributes this to the wider tolerances to temperature changes that nonmarine fishes exhibit when compared to the more equable living conditions of the tropical marine habitat. Indeed, it is shallow-water marine organisms, especially those floating near the surface or living exposed on the sea bottom, that were most severely affected by the extinctions. Deeper-water-dwelling species, such as glass sponges (which still today prefer the icy waters of the oceanic abyss), were the ones that tended to survive. All in all, as McGhee says, signs point to a rather dramatic drop in world temperature. Here, in the Upper Devonian, the very nature of the pattern of extinction seems to be pointing a steady finger at the under-lying cause, a subject to which we return after completing our survey of the major extinctions of the geological past.

The Paleozoic–Mesozoic Boundary: The Greatest Extinction of Them All (So Far)

We're back to where we began our review of mass extinctions of the geological past: the top of the Permian Period, the end of the Paleozoic. This granddaddy of all extinctions marks the dividing line between the first 325 million years of complex life on earth and the dawn of life's middle age: the Mesozoic Era, forever linked in everyone's mind with the rise and fall of the dinosaurs.

Most conspicuous in their total disappearance were the tropic coral reefs and their rich assembly of associated creatures living in and around them. Not only were the highly modified, even bizarre brachiopods (whose adaptations bespeak long pe-riods of relatively uneventful times) completely wiped out but so were the very elements that provided the structural housing of their habitat: the corals that form the building blocks of reef systems. Once established and diversified by Mid-Ordivician times, the tabulate and rugose corals (along with the coral-like sponges, the stromatoporoids) were responsible for all the reef building throughout the post-Cambrian Paleozoic.

Rugose corals generally lived as single individuals, though some grew together in true colonial fashion. Often called horn corals, rugosans were simple, generally slightly curved cone shapes that were usually 5 or 6 inches long, but could attain lengths of several feet in some species. Tabulates, in contrast, were exclusively colonial, forming large coralline heads remi-niscent of species of brain coral alive today. Some paleontolo-gists are inclined to think that tabulates were more closely related to sponges than to true corals, though ecologically tabulates filled roles we generally associate with corals.

In any case, cut back but not eliminated near the end of the Devonian, tabulates and rugosans first came back to build rather modest little reef-oid mounds (bioherms) in Lower Carboniferous times. It took a while for truly massive tropical reef systems to reappear, as they had unquestionably done by the Middle Permian. But, once again, mass extinction knocked tropical reef systems for a loop—and this time, forever—in terms of the players in the tropical reef ecological arena: Modern reefs are constructed from very different corals. Stromatoporoids persist into the Mesozoic (indeed, modern stromatoporoids have recently been identified living cryptic existences in modern reefs), but never formed the massive reefs to anywhere near the extent that they had in the Paleozoic.

Corals, and their close associates in tropical systems, were hardly the only organisms to feel the devastating effects of the Permo-Triassic extinctions. Jack Sepkoski's compilations reveal that an astonishing 54 percent of all marine families became extinct; 83 percent of the genera disappeared, and, as I have mentioned several times already, Sepkoski and Dave Raup think these numbers could translate into as many as 90 to 96 percent of all species living in the Upper Permian falling under the extinction scythe. Perhaps hardest hit of all were the very components of seafloor communities that exploded so markedly in the Middle Ordovician—groups, in addition to the corals, that have calcium carbonate skeletons and sit rooted to the seafloor quietly filtering nutrients from the ambient seawater. These include especially brachiopods, bryozoans, and several groups of crinoid echinoderms (sea lilies).

But also hard-hit were single-celled, shelled organisms (foraminiferans, or just forams). For good reason, we tend to think of single-celled creatures as microscopic. Yet there have been times when shelled single-celled organisms, especially amoeba relatives, have reached truly, if relatively, gargantuan sizes. The pyramids in Egypt are constructed of blocks of Eocene limestone, many of which consist predominantly of flat, circular forams: nummulites, meaning, "coin stones." Some of these "coins" are half-dollar size. Close relatives are still found in some tropical regions today.

Biologist John J. Lee has shown that the intricate chambers that subdivide the innards of these big shells house algae. The algae photosynthesize and presumably contribute to the welfare of the host foram that provides their housing. The algae are diatoms, normally encased in their own siliceous shells as they lead their free-living existence floating in seawater. But when they take up residence inside the shelled amoeba, they can and do dispense with their own shells.

The fusulinids were the Paleozoic equivalents of nummulites. Instead of a coiled, circular, flat shape, fusulinids wrapped themselves up like jelly rolls, looking like miniature stogies. Some were 2 inches long. They, too, had elaborate systems of internal chambers, and the betting is that algal symbiosis was also the secret of their ability to secrete such huge shells. Common in many marine environmental settings throughout the Upper Paleozoic, fusulinids failed to make it into the Mesozoic.

Trilobites, too, finally bit the dust. The youngest ones known come from the upper reaches of Mt. Everest (dramatic evidence of the power of crustal forces that can take old sea-floors to such heights!). But trilobites had already been reduced to but a few species as the Paleozoic was drawing to a close. The Frasnian extinctions had reduced them to a single order, and although they proliferated a bit in several minor radiations in the Carboniferous, trilobites never came close to the anatomical or taxonomic diversity that they had reached earlier in the Paleozoic.

This is not to say that future trilobite resurrection in a dredge haul from the abyssal plains is a complete impossibility. Stranger things have happened. Monoplacophorans are primitive mollusks that have been known from the Paleozoic since the earliest days of organized paleontological research. Their shells are simple, single, and usually rather cap-shaped—superficially something like a modern limpet. But limpets are highly modified snails. Internally, limpet shells have horseshoe-shaped markings where the main muscle mass is attached. Monoplacophorans, in striking contrast, have a series of paired muscle scars—vestiges of early days in molluskan history when there were still distinct elements of segmentation to the body. Mollusks are close relatives of the obviously segmented annelid worms and arthropods, so it is no surprise that the most primitive mollusks retain a trace of their ancient segmented patrimony.

Monoplacophorans were missing and presumed extinct after the Middle Devonian. Or they were, that is, until the Danish *Galathea* expedition dredged living specimens from the seafloor in the 1950s. We now know of several species of *Neopilina* from the abyss and the tops of oceanic seamounts. The first specimens were dubbed *Neopilina* because of their close resemblance to the Silurian *Pilina*, which, rather ironically, was a shallow-water reef dweller!

So, apparent extinction—the loss of an entire large division of life (monoplacophorans are considered an entire class of the Phylum Mollusca)—may not always be extinction. Apart from letting us hope that trilobites might someday themselves be found lurking somewhere in the abyss, the phenomenon becomes acutely important when assessing the pattern of mass extinction generally. In no situation is this more clearly the case than in the interpretation of the events surrounding the mass extinction at the end of the Paleozoic. For it is here that we confront head-on the phenomenon of Lazarus taxa.

Because of the immense spans of time paleontologists must consider, there is an inevitable fragmentation of their subject matter. Until recently, Cambrian trilobite people seldom crossed paths with their fellow post-Cambrian trilobite colleagues. The reason: It has always been difficult to see the evolutionary connections between Cambrian trilobites and those that came later. (Modern workers are having another go at bridging the gap with some success.) Because geological boundaries reflect extinction events and subsequent rejuvenations of faunas of various degrees of remote resemblance to their predecessors, they often demarcate professional boundaries as well.

Roger Batten is a Paleozoic paleontologist, a gastropod (snail) specialist trained under the great J. B. Knight. As a Columbia graduate student, he was also under the guidance of Norman D. Newell, dean of American paleontology these days, and for all times dean of extinction studies. It was Newell who virtually alone kept the study of extinction going in the United States in the third quarter of this century, and in so doing he raised the study of extinction to genuine scientific standards.

Batten did his Ph.D. thesis in the 1950s on some of the vast array of snails found living in and around the huge Permian reef complex of western Texas. Much of the rest of his career has been spent in the analysis and description of Permian snail faunas from around the world, and in trying to link up evolutionary lineages between his Permian snails and those found in older rocks of the Middle and Upper Paleozoic. Batten also looked at Triassic faunas. The Mesozoic has always been seen as the dawning of the age of modern faunas; modern snails have only the dimmest of beginnings in the Paleozoic. It is far easier to look to the Mesozoic to find obvious precursors to today's gastropods. So the Mesozoic, as a swing era in the history of life, has tended to be looked upon much more as a source of the modern fauna than it has been as a blend of truly ancient (Paleozoic) and modern (Cenozoic) life.

Early Triassic faunas tend to be dominated by ammonites. Gastropods and other forms of sea-bottom life are rather tough to find. Snails, clams, and the rest of the "normal"-looking components of seabottom communities begin to reappear in the Middle and Upper Triassic. Much to his surprise, Batten found, when surveying the old monographs on Triassic faunas, that many of the Triassic snails were Paleozoic in aspect. They looked, in other words, as though they could have been excavated from Permian, or even earlier, rocks. Some, like the genus *Worthenia*, with many species found throughout the entire Upper Paleozoic (and with close ties with older snails right back to the Ordovician), were downright Paleozoic.

Adding to the strangeness of the discovery was the fact that, toward the very end of the Permian, there are very few gastropods to be seen at all (as we shall see in a moment, there are very few, if really any, normal marine rocks of the uppermost Permian yet to be found). Paleozoic-like snails are simply more characteristic of the Triassic than they are of the Upper Permian. David Jablonski, who has already figured so heavily in the tale of understanding mass extinctions, picked up on Batten's theme, and dubbed those taxa as Lazarus taxa that seem to disappear only to show up alive and well some time after their putative demise.

Lazarus taxa tell us at least two things. First, of course, that the taxa managed to survive, even though the straightforward reading of the rocks as preserved would indicate that they had, in fact, become extinct. There are, in other words, places otherwise not yet recorded where these organisms managed to live unaffected by the extinction event. There are refugia, havens where some (by no means all!) of the creatures that appear to have become extinct in fact managed to hang on and survive. The second point is more subtle, and relies to some extent on the first: We must check those rocks where the extinction supposedly is recorded. It has been known for years that there is something weird about the uppermost Permian marine deposits as classically developed in Texas and many other areas. Salt deposits, for example, overlie the ancient reefs in western Texas. Though we can hope to glean some ideas about what caused the extinctions from such evidence, we must also realize that in many places marine conditions simply were not right for us to expect to find the usual kinds of sea-bottom-dwelling invertebrates.

This last factor makes it difficult to be absolutely sure about the actual patterns of disappearance of taxa. Only recently, especially in China, have exposures been studied in detail that suggest a continuous, uninterrupted sequence of sediments accumulating from the Upper Permian into the Lower Triassic. Absence of good continuous records that preserve normal marine environments from the Permian to the Triassic have made it very hard to be certain about the abruptness of the most massive of extinctions ever recorded. For example, in two recent reviews, W. D. Maxwell reports "complete marine records" across the boundary not only in China but also in India, Pakistan, Iran, and eastern Greenland. But paleontologist Douglas Irwin is far more cautious, claiming that the completeness of many of these exposures is dubious: There may be time missing from the rocks just when it is most critically needed to record the passage from Permian to Triassic, from Paleozoic to Mesozoic.

What we have is a picture of hordes of different taxa disappearing, but no certainty as to how abrupt the event itself actually was. Nor have we even recounted all the cast of ecological characters who left the scene, evidently for good, as they show no Lazarus effects in later rocks (and have not been found lurking in the deep sea, either). Take what happened to one group that did, in fact, survive the Permo-Triassic extinction: the ammonoids, who were shelled relatives of squid and octopi (cephalopod mollusks), and were especially closely related to the modern chambered nautilus. Like living nautilus, the ammonoids lived inside a chambered, spirally coiled shell. They were able to swim by a form of jet propulsion by expelling water with great force from a jet nozzle that they could point to change direction. Like all other cephalopods, the ammonoids were active predators of the sea.

Ammonoids, almost as much as dinosaurs, are a symbol of the Mesozoic. Ammonoids were nearly driven to extinction at the end of the Triassic, only to squeak through, radiate into a true explosion of species in the Jurassic, bloom throughout the Cretaceous—only to be wiped out, once and for all, at the end of the Cretaceous. They barely got out of the Paleozoic: Fewer than five genera seem to have gotten through, the only messengers representing the genetic information wrapped up in a vast array of Paleozoic ammonoids. Of these, one single genus surviving the debacle apparently gave rise to the vast majority of ammonoid species in the Triassic. Ammonoids had arisen in the Lower Devonian, presumably from nautiloid-like ancestors. Had that one genus not gotten through the Permian crisis, the entire history of the Mesozoic marine invertebrate faunas would have been completely different. And so on. Group after group suffered, if not complete annihilation, at least drastic cutback. But how sudden was it? And did it affect life on land as well as under the waves?

Life had managed to clamber onto land during the Silurian, and by Middle and Upper Devonian times a basic set of ecosystems, staffed with the all-important plants, plus a variety of insects, spiders, and various relatives, *and* vertebrates, had been well established. Surviving the late Devonian extinctions, terrestrial floras and faunas diversified tremendously during the 100 million years of the Upper Paleozic. The great wooded swamps that have formed so much of our coal deposits were home not only to a vast array of archaic plants but also to the early terrestrial vertebrates that were still tied to water for reproduction (amphibians), and, somewhat later, to vertebrates with an egg equipped with its own sources of protection and nourishment that forever decoupled reproduction from the mandatory aquatic environment. This amniote egg was the final solution for a truly terrestrial vertebrate existence. Amniotes without hair or feathers are reptiles, and by late Paleozoic times, reptiles had become quite abundant.

The Karroo beds (so named for the South African desert where they are best developed and most intensively studied) span the Upper Paleozoic–Lower Mesozoic boundary. They are famous for the Gondwana fauna of mammal-like reptiles: one branch of early amniotes, reptiles only because they lack the hair and other features of their mammalian descendants, was especially prolific in the southern hemisphere during these times. Their bones are preserved not only in South Africa but in South America, India, and even Antarctica, remnants of that vast supercontinent of the southern hemisphere that subsequently broke apart as the Mesozoic wore on.

At first glance, the great extinction that affected marine life at the end of the Paleozoic doesn't really show up in the Karroo. There are a series of zones with characteristically different faunas, but the sequence as a whole spans the critical boundary, which is marked simply by one of the boundaries between the series of zones. The sequence is so incomplete (terrestrial floras and faunas tend to be preserved in lake deposits, and sedimentation is far less continuous in such environments than in typical marine situations) that it is even difficult to decide whether the zonal boundaries mark extinction and reproliferation events, and, if so, how abrupt those extinctions really were.

But analysis of terrestrial vertebrates on a worldwide scale does show a rather grand turnover from the Upper Permian into the Lower Triassic. Rather than a single massive extinction, it looks as if the terrestrial vertebrate fauna underwent a grand reconfiguring through a series of miniwaves of extinction. One paleontologist, Robert Sloan, has claimed that each such wave left in its wake a progressively greater percentage of mammal-like forms: an interesting possibility that might shed some light on the causes of the biotic reshuffling at the end of the Paleozoic.

Plants, too, show a marked drop in diversity during the Permian and into the Triassic. All told, life was nearly as severely affected on land as it was in the sea. But the pattern on land seems distinctly stepped: As incomplete as the sections are, and once the Lazarus taxa are duly excluded, the picture that emerges is one of a protracted, several-million-years event—one, moreover, that is smoothly gradual and progressive only in its cumulative effect. The actual pattern of extinction and replacement is a series of less-than-catastrophic extinctions, each well above the background extinction rate, but none in itself approaching the dimensions of a truly mass extinction.

The record is even more difficult to read in the sea. Down in the Delaware basin, where all those great Permian reef invertebrates are found in such profusion, the uppermost Permian rocks are heavily laden with salt. Salt precipitates from seawater only when evaporation greatly exceeds input from rain, river runoff, or even contact with normal oceanic waters, and is usually a token, as well, of arid climatic conditions. There are late Permian salt deposits in many places around the world, betokening a widespread withdrawal of normal seawater from the continents, and drastically limiting our view of events in very latest Permian times. How can we tell how abrupt the extinction was at the boundary if we don't have the right environments, with the right faunas, in place at the critical time? Even those areas with marine rocks approaching the boundary are the source of some dispute: As we have seen, there is disagreement over just how complete those exposures really are across the Permo–Triassic boundary.

The odd nature of the marine environments toward the end of the Permian, however it bedevils our fix on the actual pattern of events, undoubtedly is telling us something about what happened to the physical environment, and thus what caused that greatest of all biological turnovers. Though Chinese paleontologists in the early 1980s reported the presence of an iridium spike (as we shall soon see, the evidence often considered the trace left by extraterrestrial impact) right near the boundary in one of their more complete marine sequences, subsequent attempts to verify the presence of an iridium anomaly have failed.

The evidence seems to be mounting that Permian extinctions in the marine environments were as progressive, stepwise, and cumulatively gradual as they were in the terrestrial habitats. There are enough rocks with the appropriate faunas in them to show that declines in group after group were well on their way in Mid-and Upper Permian times. But we may never know how bad the crunch really was at the boundary, because the rocks may never tell us.

What the rocks do tell us concerns climatic changes. Widespread polar, continental glaciers in the Upper Paleozoic were gone by the Mid-Permian, and the evaporates that marked the very end of the Permian seem to imply warmth and aridity. But the story is not so simple: We still have to understand what caused the progressive declines in sea level throughout the Permian and into the Triassic. Glaciation is by far the most efficient of known causes of sea-level fluctuation. But the movement of crustal plates also figures into sea-level changes. With the final assembly of all the continents into Pangea by the latest Paleozoic, it could be that slowdown in plate movement meant the sinking of midoceanic ridges: When plate movement is rapid, growth of these ridges is greater, and water is displaced onto the continents. Slowing the spreading rate lowers the ridges, and water spills off the continents and back into the ocean basins.

There are other possible climatic effects from assembling all the continents into a single huge mass. There is more here than meets the eye. The Permo–Triassic extinction is not only the greatest of them all but it is emerging as an interesting test case in our ability to understand the fundamental causes of immense biotic turnovers. We'll have another look at the physical and biological events at the end of the Paleozoic when we confront head-on the causes of mass extinction. Having looked at life's first 3.25 billion years of recorded history, we must now complete the whirlwind survey of mass extinctions during the 245 million years of Mesozoic and Cenozoic time.

CHAPTER FIVE

Extinction, and the Rise and Vicissitudes of Modern Life

WE ARE at the beginning of the Mesozoic. One hundred and eighty million years ahead of us looms, if not the greatest extinction of all time, certainly the most famous. The Cretaceous–Tertiary (K–T, K for *kreta*, meaning chalk) boundary marks the demise of the dinosaurs, along with a host of confreres, and is the subject of some of the more lively scientific debates. It is in this terminal Mesozoic arena that the strongest case has been made for extraterrestrial impact as the trigger of mass extinction. All the attention (in the media, to be sure, but really as a distant echo of all that thunder in the halls of academe) that the extraterrestrial impact hypothesis has attracted has fostered widespread interest in the phenomenon of mass extinction in general. Coupled with the fact that our very own biota seems so imperiled these days, extinction remains very much a hot topic not only within science but among everyone with a concern for our planet's future.

Mike Benton works on terrestrial vertebrates. He has devoted a great deal of time and attention to a group of Triassic reptiles that have languished in obscurity compared with the fame of their not-too-remote kin: the dinosaurs. Benton's reptilian crew, rhynchosaurs, have several interesting things going for them. For one, they are still very much with us, albeit in drastically reduced numbers. Anatomists of the last century recognized that, in the New Zealand lizardlike creature dubbed the tuatara, they had on their hands a genuine living fossil. Tuataras are rhynchocephalians, the last remnants of a much more widespread and diverse group that include Benton's early and middle Triassic rhynchosaurs.

Some paleontologists like to have a little fun today, pointing out that dinosaurs are not in fact extinct. They have birds in mind. Formerly, biologists recognized five great classes of vertebrates: fishes, amphibians, reptiles, birds, and mammals. Once considered discrete (if evolutionarily interconnected) branches of vertebrate existence, modern science looks askance upon these archaic divisions. Fishes, extinct and modern, are a hodge-podge of aquatic vertebrates, some (such as lungfish, the coelacanths [including *Latimeria*, another living fossil], and especially the extinct rhipidistians) more closely related to terrestrial vertebrates than they are to other fish. Amphibians are another matter: Many Paleozoic and early Mesozoic amphibians are really ancient tetrapod (four-legged) vertebrates that simply lack the advanced evolutionary acquisitions of later descendants, especially the aforementioned amniote egg. Modern amphibians (mostly frogs and salamanders), plus some Paleozoic kin, do in fact form a coherent evolutionary group. But lack of defining features is hardly a reason to consider many of the ancient tetrapods as amphibians—the more so since some are more closely related to reptiles than others.

We have already encountered the difficulty in seeing reptiles as a unified group. Rather, in addition to turtles and extinct kith and kin, there seems to have been two separate lineages of reptiles. One includes the early progenitors of mammals. The other includes, as advanced members, the birds. Birds appear to be derived from the branch that also includes snakes, lizards, crocodiles and, most especially, dinosaurs. The latest and probably very good idea is that birds sprang from somewhere in and around the carnivorous, theropod dinosaurs, meaning *Tyrannosaurus* and the like.

So birds are dinosaurs with feathers, or carnosaurs without

teeth, but they hardly qualify as living fossils. Not only are birds as diverse as dinosaurs ever were but they are not at all the same sort of thing as even a highly active, small, agile carnivorous theropod was. Ecologically, birds aren't dinosaurs. Ecologically, tuataras fall into the spectrum of behaviors inferred for their Mesozoic kin.

Benton has done much to establish rhynchosaurs as the ecological counterparts of the herbivorous dinosaurs that were soon to replace them. Rhynchosaurs were vegetarians that radiated into a variety of diverse forms, filling niches that in Benton's view had been most recently held by herbivorous mammal-like reptiles (who nonetheless continued as prominent members of the terrestrial ecosystem throughout the middle Triassic). Thecodonts, also in the dinosaur/bird (as opposed to mammal) side of early reptiles, shared the carnivore slots with mammal-like reptiles in the middle Triassic. The difference between thecodonts and dinosaurs is largely technical: Most of us would call most thecodonts small dinosaurs.

Then, abruptly, it was all over. After the first stage of the upper Triassic (the Carnian), the two major clans that together comprise the dinosaurs—the saurischians and ornithischians—became the ones to staff the terrestrial ecosystems. Sauropods are the big, lumbering, herbivorous giants (*Brontosaurus*—which, as any kid will tell you, is properly called *Apatasaurus*); together with the carnivorous theropods (e.g., *Tyrannosaurus*), they form the Saurischia. Ornithischians include duck-bills, ceratopsians (such as *Triceratops*), stegosaurs, and ankylosaurs; all were herbivorous.

Once again, though, the pattern is a bit difficult to read. For example, the muds and muddy sands of the Karroo beds, with all their mammal-like reptiles, are overlain by Upper Triassic sandstones with a few mammal-like reptiles and a bunch of newcomers—dinosaurs, among the oldest yet discovered. The first inclination would be to invoke an extinction, and claim that dinosaurs evolved to take the place of those fallen soldiers of the past ecological regime. But paleontologists had to wonder where the dinosaurs actually came from (obviously not springing directly from creatures now preserved in the immediately underlying beds). And, as Benton himself has pointed out, the recent tendency to reinterpret dinosaurs with greater respect—realizing, in other words, that dinosaurs themselves were no slouches but fully active and able holders of their ecological domains—led to the speculation that dinosaurs in fact competed with, and eventually defeated, the denizens of the Middle Triassic ecological world.

But dinosaurs had their undeservedly low reputation because of their supposed loss in the battle for ecological space with none other than ourselves: mammals. We now know otherwise: Dinosaurs did just fine, filling ecological space the world over for 145 million years, yielding just enough space for us mammals to cling to. Mammals arose at just about the same time that dinosaurs got going (the oldest known mammals are Upper Triassic). It is now abundantly obvious that it took the eradication of dinosaurs to give mammals a chance to take over, which we did, but not until after the great K–T extinction.

So we have had to abandon the notion that mammals took over from dinosaurs (and other, related denizens of the Mesozoic landscape) by sheer dint of inherent superiority. We were lucky to have hung on long enough to get the chance to inherit the earth. That being the case, it makes no sense to paint a picture of dinosaur superiority gradually evicting rhynchosaurs, thecodonts, and mammal-like reptiles from their Mid-Triassic hold on ecospace. Rather, Benton suggests (and I concur), that the evidence points to extinction of the old that paved the way for new inheritors of the earth. Dinosaurs were simply survivors. They were a small fraction of the terrestrial fauna when the earlier fauna was still intact in the Carnian, the earliest part of the Upper Triassic. It was the dinosaurs who were able to beat everyone back when it came time to set up ecological shop once again during the Norian, the latest stage of the Triassic.

Once again, though, it is difficult to be precise about the

timing, and thus the exact nature, of extinction events in the latest Triassic. There is no question that, as a turnover event in the history of life, the Upper Triassic ranks right in there with the Upper Ordovician, late Devonian, end Paleozoic, and the K–T. Benton, citing data compiled by Sepkoski, himself, and others, reports that more than 20 percent of known families became extinct. In the seas, sponges, snails, clams, and brachiopods were particularly hard-hit. Once again, the ammonites (who barely squeaked across the Permo–Triassic boundary) very nearly came to the end of the line. Ceratites had radiated explosively and utterly dominated Triassic ammonite communities; they disappeared without a trace at the period's end: It was one of the several low-level diversity lineages, themselves Paleozoic holdovers, that survived and produced most of the last great ammonite radiation of the Jurassic and Cretaceous.

Once again, it appears that extinctions toward the end of the Triassic came in several waves. Though there was a peak in extinction right at the very end, there was another one at least some 10 million years earlier. A minimum of two extinction peaks on both land and sea seems to be the emerging consensus, though the extent to which events on land and sea were actually synchronous remains unclear. In other words, as much as 10 million or perhaps even 20 million years may be involved in the dismantling of early Mesozoic biotas, and the founding and rise of succeeding Jurassic and Cretaceous biotal systems.

The best evidence for a single, abrupt blow underlying a mass extinction is yet to come, but we should not forget that in singling out a period of time as long as 20 million years for an extinction event is by no means to imply a regular, steady, gradual decline in ecological diversity. No such pattern is implied. But the mass biotic turnover is indeed cumulative, forged from more than one distinct episode where extinction rates vastly exceed the background norm. Nor should we lose sight of the fact that, though protracted, sometimes over several millions of years, such massive turnovers of the biota still are way out of the ordinary, following as they do far longer periods of status quo, business as usual with only background rates not only of extinction but of evolution as well. The greater the mass extinction, regardless of how it may be caused, or how long it takes to run its course, the greater the effect it has in shaping the future evolution of life on earth. Consequently, the differences between the old and new ecosystems will be more profound.

The Plot Thickens: Patterns in the K–T Mass Extinction

Walter Alvarez was a young geologist at Berkeley when, in 1977, he decided to do a detailed analysis of the sediments across the Cretaceous–Tertiary boundary. Interest had already begun to rise in the events at the end of the Cretaceous, when the Mesozoic ended, the Cenozoic began, and the dinosaurs disappeared once and for all, paving the way for the rise and radiation of mammals.

Alvarez needed to study rocks that held as complete a record of events as could be found anywhere: exposures of rock formed by continuous deposition of sediments from the Upper Cretaceous well into the lower Tertiary. The uppermost stage of Cretaceous time is called the Maastrichtian, named after a town in the Netherlands; the succeeding stage and the lowest division of the Paleocene (itself the earliest segment of Tertiary time) is the Danian, named after the exposures in Denmark. Stevn's Klint, where we began this odyssey of ancient extinctions and their connections to modern, troubled biotic times, is a classic place to study the transition from Maastrichtian chalks up into the Danian chalks and limestones. But geologists had long known of the subtle signs that there is time missing at the K–T boundary at Stevn's Klint. The slightly wavy line that separates Cretaceous from Tertiary along the cliff face hints of a hard-bottom seafloor, one that was not collecting sediment for some unspecified length of time near the Cretaceous–Tertiary boundary.

Alvarez knew that one sequence with a more regular and possibly continuous accumulation of seafloor sediments is exposed on the hillsides near Gubbio, Italy. Carefully sampling at short intervals up the exposure, Alvarez was particularly struck by a half-inch-thick band of red clay that seemed to occur right at the Cretaceous–Tertiary boundary. He carefully collected samples on either side of the boundary and brought them back to California for further study.

Walter Alvarez's father, Luis, was a Nobel prize–winning physicist who had, by that time, already enjoyed a long and rather varied career. Among other things, the elder Alvarez had figured importantly in the assembly of the atomic bombs that had wrought such catastrophe and brought such a sudden end to World War II. It was Luis who suggested an analysis of the atomic composition of the red clay sample that son Walter had brought back to Berkeley. The idea was as simple as it was ingenious: Trace elements, such as the metallic element iridium, accumulate in minute amounts as part of the steady rain of cosmic dust that is constantly filtering through earth's atmosphere and settling on the surface. The extremely slow rate of accumulation of this meteoritic dust has been calculated. Measuring the amount of iridium gives a clue of what percentage of the grains of sediments came from outer space. The Alvarezes planned to use that information to calculate how long it took for the sediments derived from normal, earthbound sources to accumulate. In other words, measuring the percentage of iridium in the half-inch-thick layer of red clay at Gubbio would, in principle, reveal how much time it took for that red clay layer to form.

But, when they and their team at Berkeley actually analyzed the iridium content of the red clay, they stumbled on something completely unexpected. Their analysis made the headlines: The rare metallic element iridium (allied with platinum and just as scarce and valuable) was present in the red clay samples at levels thirty or more times the normal occurrence of iridium in sedimentary rocks and soils, or for that matter, in any earthly crustal materials save some volcanic rocks. Such concentrations otherwise were to be expected only from deep within the earth. But, significantly, such concentrations are also known from meteorites. Comets and meteors (bolides— visitors from space, within our solar system but far beyond the earth's own orbit) could therefore be expected to house iridium in such high concentrations. The Alvarez team concluded that by far the most likely source of the iridium was contact with an asteroid or comet—a visitor from outer space. And then there was the clay itself to be considered, sandwiched as it was between limestone beds. Where had all those clay particles come from, anyway?

The Alvarez team drew up a scenario that saw a collision of the earth with a huge extraterrestrial object, with resulting clouds of pulverized dust encircling the earth, occluding sunlight and impeding photosynthesis, thus triggering massive, global ecosystem collapse overnight. Their picture gained instant attention, at first among a startlingly diverse array of scientists, and eventually in the public media. Physicists, geochemists, and astrophysicists began to tackle mass extinction, which was once the sole purview of geologists and paleontologists. Astrophysicists jumped in because of the suggestions, emanating from Dave Raup and Jack Sepkoski, that mass extinctions were periodic: cyclical events with a period of some 26 million years. Such regularity virtually begged for astronomic causes, assuming, that is, that mass extinctions are triggered by collisions between the earth and extraterrestrial bodies.

Right away, of course, there were fights: Some paleontologists insisted that the data at the end of the Cretaceous showed gradual extinction over one, or even several, millions of years— not the overnight wipeout needed to fit the Alvarez picture. Others felt that the iridium must have come from volcanoes, not meteorites. The wrangling goes on; though, as we shall see, some clear patterns are indeed emerging. One thing is abundantly clear: The high concentrations of iridium at Gubbio are not only for real but they are no unique fluke. Walter Alvarez went right out to Stevn's Klint and verified high levels of iridium in the fish clay. Others reported similar iridium spikes (high blips on an analytic graph, showing dense concentrations) from many other localities. Still others have found grains of shocked quartz—telltale evidence of tremendous pressures generated when meteors or comets strike the earth.

There is no doubt that some dramatic event, possibly volcanic, but more likely extraterrestrial, did occur right at the end of the Cretaceous. But was it a single event, and did it single-handedly do everything in, or was the Cretaceous extinction a more complicated matter? How do these apparent events figure in as a contributing cause (or even the root cause) of the mass extinction at the end of the Mesozoic? The first issue to be resolved is one of rates: How rapid were those Upper Cretaceous extinctions? To get at this issue, we must turn to the rocks and fossils themselves, as interpreted by the many skilled paleontologists who have focused their attention so closely on the Upper Cretaceous, especially since Walter Alvarez brought the iridium anomaly at the boundary to the surface of our consciousness. Then we can see how the extraterrestrial scenarios fit in with an emerging picture of what actually underlies mass extinctions in general.

It was the Alvarez team, with its dramatic pronouncements, its linking of geochemistry with paleontology, its Nobel-laureate credentials, and its sheer exuberant passion that catapulted mass extinction into big-time media. That, in itself, was a good thing: The more we are aware of events in the geologic past, the better able we will be to face our current ecological problems. This is so, even if the only lesson we get from the past is the conviction that extinctions, truly devastating, global mass extinctions, are real. But one possible drawback to the extraterrestrial scenario might be a relaxation of present concerns: If it takes collisions with comets to kill nearly everything off, current events, and especially our own human contributions, might be passed off as in no way a threat of comparable dimensions. And that would be to misjudge the import of all past extinctions, including the K–T saga, regardless of its ultimate cause.

There is much more to the extraterrestrial story: sorts of events, what causes them, and just exactly how extraterrestrial events may conspire to cause mass extinctions. They are the subject of the next chapter, after we complete our survey of the actual extinction patterns themselves. But, lest anyone get the idea that extraterrestrial events are the very deus ex machina that gets us off the hook as potential causal agents of the next big extinction event, be it known that those agents from outer space simply trigger ecosystem collapse in ways that are fairly mundane and, by now, familiar.

There is a strong school of thought that sees extraterrestrial impacts more as a coup de grace than the prime mover in mass extinction: As much as geologists and paleontologists have focused an inordinate amount of effort on events in the latest Cretaceous, we are still unsure about some very fundamental facts. Though many have firm opinions, as a group there is little unanimity on the actual patterns of terminal Cretaceous extinction.

Late Cretaceous Life and Death on Land and in the Sea

It was really after the late Triassic extinctions that life began to assume a modern complexion. True, ecosystems were to continue to be dominated by dinosaurs on land, pterosaurs in the air, and a number of now-extinct reptiles and invertebrates (especially the ammonites) in the sea. But as the Jurassic wore on and gave way to the Cretaceous, the angiosperms (modern seed plants) became much more diverse. Modern bony fishes (teleosts) also arose in the Upper Jurassic, continuing their expansion through the Cretaceous. Among invertebrates, modern forms of snails and clams also began to come into prominence in the Mid- and late Mesozoic. The Mesozoic really is aptly named: There is a definite intermediacy between the truly archaic aspect of the Paleozoic, and the modern, or very nearly so, appearance of fossils of the Cenozoic. To a remarkable degree, it takes an expert to tell 30-million-year-old snails and clams from the species we have living in today's oceans.

But we want to know what happened as the end of the Cretaceous neared. Just how abrupt were the disappearances of species, and how pervasive were the extinctions? Sure, all non-bird dinosaurs—constituting a major chunk of the diapsid reptiles and being the major vertebrate players in the varied terrestrial ecosystems for 145 million years—disappeared forever. But were the dinosaurs cut off in their prime? Or were they like trilobites that departed this earth more with a whimper than a bang, reduced as their numbers were to but a few species as the Paleozoic came to a close?

Western North America is a gold mine of Jurassic and Cretaceous terrestrial faunas: dinosaurs aplenty, of course–as everyone who has ever visited a natural history museum in the United States and Canada knows full well. But mammals, too. Critical especially to the claims of an abrupt, apparently extraterrestrially caused event, plants as well are abundantly represented in the thick layers of nonmarine sediment exposed in the old basins in and around the Rocky Mountains. Montana and southern Alberta are particularly important here, as these are the areas that preserve many exposures of the transition from Cretaceous to Tertiary in terrestrial ecosystems.

Paleontologist Robert Sloan has carefully analyzed the diversity of dinosaur genera in the latest Cretaceous. The pattern shows a standing crop of diversity of approximately thirty genera from around 76 to 70 million years ago followed by a progressive decline: In western North America, we know of twenty-two dinosaur genera 69 million years ago, eighteen around 68 million years ago, twelve some 67 million years ago, and only seven when we reach 66.7 million years ago—the date that Sloan and his colleagues use for the Cretaceous–Tertiary boundary. These dates do not correspond exactly with the standard numbers used in the latest (1989) internationally agreed-upon chart of geological time—the compilation that I am following in this book. The date given on that chart is itself equivocal, as both 65 and 64.6 million years ago appear, reflecting ongoing study and refinement. Radiometric dates are of course literally not written in stone: They are calculated from hundreds of replicated measurements of radioactive decay of a variety of isotopes, subject to a certain degree of error, and must, furthermore, be extrapolated to fossil-bearing beds that cannot be directly dated radiochemically.

But regardless of the status of the dates that Sloan reports, the pattern is very clear, at least for western North America. Dinosaurs were on the decline in these particular ecosystems for the last 3 million years or so of Cretaceous time. The decline appears to have accelerated as the end of the Mesozoic drew near. But against the scenario of gradual decline there is the equally important observation that a minimum of seven dinosaur genera (twelve known species, including the renowned *Tyrannosaurus rex* and *Triceratops horridus*) seem to have been present right at the boundary. And though Sloan has argued that, in some places, they managed to squeak across the boundary (a position hotly disputed by many experts), there is little doubt that the dinosaurs had not suffered their own intrinsic decline and fall completely before the end of the Cretaceous.

Then there is the matter of those reptiles that did make it across the boundary pretty much unscathed. Snakes, in the form of boas, slithered across, and none of the families of both lizards and turtle alive in the Maastrichtian died out at the K–T boundary. The same is true of crocodiles, superficially dinosaurlike. The puzzle has always been: What is it about these groups (all of which, of course, have representatives still very much alive today) that allowed them to get through the event with only minor extinctions, while the dinosaurs and the last remaining family of pterosaurs (flying reptiles) didn't?

Jablonski's distinction between background and mass extinction may be useful here, though the difference is predicated on the notion that mass extinction is such a coarse sieve that ecological differences, mirrored in the different adaptations of various creatures, are less significant in determining which groups survive and which don't make it through. Cretaceous patterns are beginning to look a bit like others we have already encountered: Habitats do seem to make a difference. Aquatic organisms, for example, seem to be less affected than their dry-land counterparts when patterns of mass extinction in terrestrial (i.e., as opposed to strictly marine) biotas are examined. This may help explain why turtle, crocodiles, and even the boid snakes survived while other lineages did not. Many of our living boids readily take to water, not just the anacondas of South America but even the huge reticulated python of Africa, which often hunts submerged in lakes.

But the mystery is not entirely dispelled by this argument: It is still not at all clear why lizards and rhynchocephalians (so dinosaurlike in their terrestrial habits and habitats) made it while the dinosaurs failed to hang on. Nor is it a matter of size, with only large-body dinosaurs succumbing while the smaller lizards and rhynchocephalians evaded the reaper. There are large relatives of the Komodo dragon (largest and nastiest of modern lizards) known from the Cretaceous; and, of course, not all dinosaurs were huge (though most that survived to the bitter end of the Cretaceous were of the more conventionally large dinosaurian dimensions). The effect of body size on extinction emerges as an intriguing theme when we look at patterns of mammal extinction during the Ice Ages. We note it here, though, as it will help sort out competing scenarios of what is really underlying bouts of mass extinction.

In the sea, it's somewhat the same thing: There is a theme of relative protection, based on exactly where organisms are living, that seems to be correlated to a great degree with probability of survival in a mass extinction. Microscopic, floating organisms living near the oceanic surface are often hit, suddenly and hard, during extinction events. Life on the seafloor, though, is almost invariably less severely and dramatically affected. And, among those bottom dwellers, it is those in relatively shallow waters, such as the seaways encroaching into continental interiors, that are at highest risk. And even in those situations, burrowers who live within muds and sands of the seafloor seem to fare better than their less well-protected cohorts swimming over, reclining on, or attached to, the sea bottom.

Chalk, so prevalent in the Upper Cretaceous, is primarily composed of unimaginably huge numbers of ultra-tiny plates—parts of the skeletal system of single-celled, photosynthesizing algae. These are the coccolithophores. Single-celled photosynthesizing algae form the very base of the food chain of the ocean, the underpinnings of the vast bulk of biological energetics of the world's oceans. It is fortunate indeed that in the Jurassic a number of different elements of the oceanic plankton system developed hard, skeletal features: Not only the coccoliths but also floating, carbonate-shelled amoebae (the foraminiferans whose distant, seafloor-dwelling relatives, the fusulinids, figured so heavily in events in the Upper Paleozoic) first appeared in the Jurassic. This is a minimum date: Because the sea bottom is in a constant state of flux, with new seafloor continually produced at the midocean ridges, and old seafloor, at plate edges, being pulled down into trenches and sliding under continental crust to disappear forever, the oldest oceanic crust still in existence itself is only Jurassic in age. But planktonic forams and coccoliths do show up in near-shore sedimentary deposits from the Jurassic on, and we can be pretty sure that calcareous plankton really did develop sometime during the Jurassic Period.

In 1965, in *Science* magazine, oceanographic paleontologist M. L. Bramlette published a striking diagram documenting the extinction of calcareous algae right at the Cretaceous–Tertiary boundary. Bramlette's graph showed that twenty-seven of the twenty-eight known species marched right up to the boundary, then disappeared forever. Only one species was definitely shown to have made it across. But right at the base of the Danian, there was a great efflorescence of new species: Species-level turnover was abrupt, severe, and dramatic.

Bramlette's data are important for several reasons: He was dealing with the base of the oceanic food chain, the data reveal strikingly abrupt extinction, and coccoliths, as members of the phytoplankton, are living at or quite near the oceanic surface, presumably significant when it comes to establishing the cause of their demise.

In an extremely significant study at the Tunisian locality known as El Kef, paleontologist G. Keller studied not the coccoliths but the planktonic foraminiferans, a step or two up in the food web. Keller sampled in very closely spaced intervals below and above the K–T boundary, which was taken to be the thin layer with the iridium anomaly, corresponding to those documented by the Alvarez team at Gubbio and later found at Stevn's Klint and many other places where the K–T transition is exposed.

Superficially, Keller's foram data look exactly like Bramlette's coccoliths: abrupt extinction of many Cretaceous species, followed just about as abruptly by a formidable radiation of new species right after the event. But Keller's detailed, picky sampling revealed a more complex story—one that has direct implications for the interpretation of the K–T extinctions as a single, virtually overnight episode. Rather, Keller's forams became extinct in several distinct pulses, themselves rather abrupt, but amounting to a complex and rather protracted composite event when taken overall.

Planktonic foram extinction at El Kef began 25 centimeters below the iridium anomaly, where six species drop out of the record. Eight species drop out right at the iridium layer. But twenty-two more drop out, in two separate waves, within 7 centimeters above the boundary! No such restudy of Bramlette's coccolith data is as yet forthcoming, but the odds are good that what really did look like a single-wave, abrupt extinction will probably turn out to be a stepwise series of lesser events both below and above the K–T boundary: After all, Bramlette's work was done before the iridium anomaly had been discovered, and there was no way the documentation of extinction patterns could have been as precise as was possible when Keller did his work on the planktonic foraminiferans.

What about the other, larger, bottom-dwelling marine invertebrates? In a compelling chart of brachiopod occurrences right around the Maastrichtian–Danian boundary in northwestern Europe, compiled by paleontologists M. B. Johansen and F. Surlyk, there seems to be an abrupt extinction of some twenty species right at the boundary (just below the boundary clay layer). Six species, interpreted as ecological generalists, do manage to get across the boundary. Twenty-three species are depicted as arising right at the base of the Danian. All in all, brachiopod extinction seems very abrupt compared with rates of background extinction for brachiopods in the Upper Cretaceous. The subsequent radiation also sets off the seeming starkness of the event.

But Surlyk and Johansen also present the detailed data for Nye Klov, after Stevn's Klint, the next most famous exposure of the K–T boundary in Denmark. Here the pattern is a bit more complex. Eight species drop out just short of the boundary, which, as Surlyk claims, might well be an accident of sampling but might, as well, represent their true extinction. Eight other species do drop out right below the boundary clay (as all twenty-three were depicted on the composite chart), while the remaining seven of the twenty-three species known to become extinct in the Maastrichtian actually appear in broken and abraded condition within the clay itself. Surlyk is probably right that the occurrence of these seven species within the clay represents the reworking of dead shells into the clay, but he admits that they might represent genuine survival.

In keeping with the usual patterns of evolutionary rebound and diversification following extinctions, it turns out that there is a barren interval of 3 or 4 meters before the Danian brachiopod species actually appear. Extinction of brachiopod species in the Maastrichtian appears to have been somewhat more stepped, and recovery somewhat slower to take place, than the composite chart of brachiopod occurrences across the boundary would seem to indicate. The difference between the composite chart and the details at Nye Klov are somewhat akin to the differences between Bramlette's coccolith chart and the foraminiferan data assembled by Keller at El Kef.

But, however complex the actual pattern of extinction and subsequent redeployment of newly evolved brachiopod species may have been, something certainly startling, and rather abrupt, terminated the bulk (80 percent of known species) of the brachiopod fauna at the K–T boundary in northwestern Europe. Interestingly enough, Surlyk points out that it is brachiopod species adapted to the unusual habitat of chalk seafloors that become extinct; the six generalist species manage to make it across. Chalk deposition ceased abruptly at the boundary, as the hard ground was formed and clay deposition commenced; it took a while for conditions suitable for true chalk formation to be restored, and it was not until then that surviving species were able to sow the seeds for a new radiation of chalk-adapted brachiopod species.

The chalk environment also makes it difficult to chart the course of extinction and evolution among other invertebrate groups, otherwise normal and abundant members of marine sea-bottom communities. Not only is the chalk environment unusually low in diversity and perhaps even in numbers of organisms present—making fossils difficult to find—but apparently those whose shells are composed of the more stable form of calcium carbonate (calcite rather than aragonite) are preferentially preserved. This makes trouble for molluskan paleontologists trying to understand what happened to snails, clams, and ammonites. The evidence that is there does show significant species-level extinction of bivalves (clams) at the boundary, but also that a number of species typical of the Tertiary had already appeared in uppermost Cretaceous times. Thus opinions differ, with some paleontologists opting for an interpretation of bivalve change as gradual at the K–T boundary, whereas others opt for a more abrupt turnover.

One aspect of bivalve diversity that leads to notions of gradual decline is the very obvious loss of large and distinctly Mesozoic forms of bivalve life. Inoceramids were large flat-shelled clams that simply lay there on the seafloor much in the way oysters do (and were doing back then as well). Some inoceramids measured 4 feet across, and their numbers as well as their bulk make them common elements and sure signposts of Jurassic and Cretaceous sea-bottom communities. Surlyk reports that inoceramids had dwindled severely by Upper Cretaceous times: Four species did make it into the Maastrichtian, but none made it all the way to the end.

Even more unusual, and utterly characteristic of Cretaceous clam life, were the rudists. Like the rhicthofenid brachiopods of the Permian reef communities, rudistids adopted a shell form uncannily like a coral. The bottom shell was lengthened into a deep cone, and the upper shell served simply as a flapping lid. The lid opened when conditions were right for feeding and respiration, and closed for protection when the animal was resting or under attack. And, sure enough, these coral-like clams went right ahead and built thickets and small reefs—a remarkable aping of habit and ecological construction of true corals. But their heyday had been in early Cretaceous times, and they were all but gone by the time the late Cretaceous extinction events occurred.

Because some of the most prominent and uniquely characteristic elements of the Mesozoic fauna got nearly to, but did not quite make, the very end of the Cretaceous, the impression is that extinctions in general at the end of the Mesozoic were, in the main, gradual. There is certainly a gradual element: Even within the Maastrichtian, the stepwise microanatomy of the pattern could be interpreted as gradual (though that would be an abuse of the word). But we can't let ourselves be misled by the fates of a few conspicuous groups: Against the gradual dwindling of some taxa we have the clear evidence of dramatic events at the very close of the Cretaceous. The ammonites (close relatives of the pearly nautilus) demonstrate this very well.

Ammonites, the very hallmark of Mesozoic life in the seas, were declining in diversity for much of the Upper Cretaceous. Paleontologist Steven M. Stanley refers to ammonite evolution as "boom and bust": Almost wiped out at the end of the Paleozoic, the great radiation of ceratite ammonites, so characteristic of the Triassic, appears to have sprung from just one of the five or so genera known to have survived the Permo–Triassic extinction. And only two genera seem to have evaded extinction in the Upper Triassic, one of which appears to have provided the basis for a tremendous radiation of ammonite taxa in the Jurassic and Cretaceous.

It took events at the K–T boundary finally to do the ammonites in. And while it is quite true that ammonite diversity had been on the wane as the Upper Cretaceous wore on, it is simply not true that they had dwindled down to a precious few (as, for example, trilobites had near the end of the Paleozoic). Some groups of ammonites were distinctly alive and well, flourishing and showing rapid rates of evolution even as the Upper Cretaceous was in its final phases. These taxa very definitely were cut off abruptly at the pass.

Ammonites are externally shelled, internally chambered relatives of squid and octopi. Their shells are generally coiled in a simple and elegantly beautiful, logarithmic spiral. From time to time in their long history, though, both ammonites and nautiloids departed from the norm, constructing straight shells, or J-shaped conchs. Some of the very latest ammonites departed from all curvilinear regularity, becoming a twisted and even knotted morass of tubing, presumably betokening a stationary life on the seafloor, abandoning active swimming habits. Some of these oddities prompted paleontologists of a bygone era to speculate about internal causes of extinction: the racial senescence to which we return in the next chapter. But what such unusually shaped organisms really imply is that, in some branches of the ammonite pedigree, adaptive evolution was very much in full swing when the Cretaceous came to a close.

So we get two signals with ammonites: Yes, their diversity had been declining. This is apparently so even though it is admittedly difficult to find ammonites in chalk deposits: They didn't preserve well in such environments, and very possibly didn't like swimming around in seaways with chalky bottom sediments. Like much of the calcareous granular sea bottoms in the Bahamas (which can effectively be biological underwater deserts, so devoid of life are they often found to be), chalk bottoms just did not support the riotous assembly of organisms that would, in turn, support many predators at, or very near, the top of the food chain.

But it is at the same time true that many species of ammonites were alive and well in the Maastrichtian, apparently until its very end. Ammonites give us both signals, and if their wipeout was not as hugely catastrophic as is sometimes supposed, there was nonetheless a strong element of sudden disappearance of what remained of a once more diverse array of creatures. The ammonite story is strongly reminiscent of the dinosaurs: They, too, had been dwindling. But something really did happen right at the end of the Cretaceous to take out the remaining species in what seems to have been an unceremoniously abrupt fashion.

We have followed a slightly tortuous path toward understanding patterns of late Cretaceous extinction. On the one hand, group after group shows a drop in diversity throughout much of the Upper Cretaceous. On the other hand, there is always some branch or other within each of these groups that seems not only to survive to the very end but in some cases is flourishing and even evolving rapidly. And, from another direction, on the one hand there are occasional studies that show

Cretaceous extinction & trends

a drastic species-level wipeout right at the K–T boundary, a cutback that is immediately followed by a reradiation springing from the one or two species that manage somehow to get across the boundary. On the other hand, detailed analysis of carefully compiled data sets seems to show that those pictures are themselves composite: Finer resolution seems to show that extinctions at the very end of the Cretaceous came in several waves, only one of which was right at the boundary (meaning the sedimentary layer, usually if not always very thin, with the elevated iridium and other rare earth elements). Some of the waves were just before that boundary layer, whereas others seem to have come after. And reproliferations, when looked at closely, invariably came after a considerable interval of time. Representing evolutionary rebounds as immediately following extinctions helps to dramatize the abruptness of the ecological turnover, and perhaps to heighten the interpretation of mass extinction as being ecologically literally instantaneous.

We seem, in other words, to be converging on a scenario of rather abrupt events terminating the Upper Cretaceous—a time when many, if not all, habitats preserved in the record were supporting an array of life already somewhat reduced in diversity. Each of the events was apparently quite abrupt in ecological time. But how much time was consumed as this series of events was going on is hard to judge: perhaps as much as a few hundred thousand years.

But the original Alvarez impact hypothesis explicitly invoked an immediate and utterly catastrophic event. The scenario was quickly expanded to incorporate months and even years. But the impact hypothesis definitely speaks in terms of true ecological time: days, months, years, or perhaps a few decades—not centuries, millennia, or fractions of millions of years.

The problem we paleontologists have in pinpointing events in ecological time has always made it difficult to translate and apply ecological processes that are known from a long history of empirical and theoretical investigations of the organization of biotic systems on the earth right now. Often, simple extrapolations of processes that take place over a few years or decades into geological time are nearly completely at odds with the patterns we can see in the paleontological record. This has certainly been the case of comparing extrapolations of evolutionary processes into geological time: For well over a century, paleontologists had been looking for gradually evolving lineages in geological time.

Instead of seeing the characteristics of organisms change continually and slowly through hundreds and thousands of millennia, we much more commonly find species persisting almost unchanged through their multimillion-year histories. One of the cornerstones of the theory of punctuated equilbria (which I originated with Stephen Jay Gould in 1972) was the recognition of this tremendous stability (we called it *stasis*) that reflects the more typical, actual evolutionary history of species rather than the extrapolations of processes studied over days, months, and years by evolutionary biologists working with modern organisms. The studies themselves are fine: The problem lies in understanding how the evolutionary process actually works through geological time.

But our problem here is a bit different: How would we be able to tell an instantaneous ecological event if we had one 65 million years ago? The iridium anomaly certainly helps: Some geological events leave widespread, often chemical, traces of short-term events. These include not only putative fallouts from meteoritic or cometary impacts but also ash-falls from volcanoes, which can cover thousands of square miles (as the Mt. St. Helens eruption did). Marine rocks often preserve volcanic ash layers, and these can often be matched up nearly across entire continents because each volcano is likely to produce a characteristic chemical profile different from any other. But some of the data just below, at, and just above the iridium layer seem to show a series of extinctions, and the degree to which we can confidently accept an ecological collapse scenario triggered by one (or, increasingly more popular, several successive) impact is still equivocal.

Plants and Ecological Havoc

Paleobotanist Garland Upchurch has carefully reviewed the criteria for recognition of catastrophic versus gradual extinctions, especially from the point of view of climate change. Many earlier paleontologists had advocated a long-term cooling of Cretaceous climates, leading to a gradual accumulation of extinct taxa. One way, as we have seen in the Permian case, to tell whether things are getting cooler, warmer, or staying pretty much the same is to evaluate the evidence for glaciation; however, as we saw, such information can be difficult to interpret. Another way, of course, is to look at the plants, providing that enough is known of what sorts of plants prefer colder climates and which are tropical in their preferences. By the end of the Cretaceous, plants were sufficiently modern in aspect (meaning the immediately recognizable ancestors of modern-day floras were sufficiently in place) so it is a relatively straightforward matter to tell the difference between cooler, high-latitude floras and those of the more temperate, subtropical and tropical realms.

But plants, as parts of the biota, are also potential victims of extinction. We have, in plants, organisms whose very patterns of relative persistence and extinction can be checked directly against their signals of climatic preference. And we need not find fossil leaves, stems, trunks, and roots to get this information from plants: Each species of complex plant has its own unique form of pollen. And pollen is easily preserved and easy to find—much easier than leaves and other portions of identifiable plant anatomy.

Upchurch states the predictions for catastrophic extinction succinctly. Foremost among them, we would expect direct evidence for a truly mass kill at the extinction boundary. And that is precisely what the floral record indicates: A number of typical Cretaceous species persist to within a few centimeters of the boundary. The boundary itself, at least in many terrestrial localities in North America, consists of a K–T boundary clay, as enriched with iridium and full of shocked minerals as its more famous counterparts in marine sediments. But, even allowing for stepwise kill-offs just before and after the iridium level in the marine record, there still is plenty of evidence that a number of marine species also disappeared right at the boundary. What is so special about these terrestrial plant data?

Ironically, what convinces Upchurch (and most of the rest of us) that there really was a mass kill of terrestrial plants at the K–T boundary is not so much the species that drop out just below the boundary clay but what happened later, in the basal Paleocene! Angiosperms dominated the floras right up to the boundary. But it is ferns that ruled right after the boundary clay, which is exactly what happens after volcanic explosions destroy angiosperm-dominated forests today. It is what happened immediately after the boundary clay, with all its evidence of sudden calamity, whether from massive volcanic eruption, or, more likely, from spectacular collision with an extraterrestrial body, that tells us how truly disrupted plants really were. Gone, for a while, were angiosperms, replaced by but a few fern species. The terrestrial ecosystems seemed to have been knocked for a loop.

Upchurch thinks we are looking at the direct result of wildfires that are best documented in North America but presumably were global in extent. Wildfires killed off a lot of plants and resulted directly in a number of species becoming extinct. Wildfires triggered an ecological upheaval that further engendered a taxic upheaval, when the kill-off was so severe that entire species disappeared. Upchurch does not think that all the damage accrued immediately upon impact: Some groups made it through for a while, only to succumb in what looks like a delayed reaction. Crippled beyond repair, extinction for some of these species came a bit later than the actual, overnight devastation of the impact itself.

Upchurch made several other predictions, that is to say, observations we would expect to make if the impact scenario is basically correct. For example, we would expect to see a gradient in effect, with higher-latitude plants doing progressively better than temperate and tropical plants. Those plants that we would expect to survive such a catastrophic event would have to be able to be dormant for a while. In higher plants, both the plant itself as well as its seeds can be dormant for varying periods of time. The higher the latitude, the greater the dormant ability that is already built in.

As ecologist George Stevens pointed out, high-latitude organisms in general are adjusted to seasonal extremes much more than their lower-latitude brethren. Upchurch's dormancy argument amounts to the same thing: The scale of susceptibility to extinction should be highest in the tropics, decreasing as one gets into the higher latitudes. Based on actual knowledge of latitude of these ancient floras, and the physiological tolerances of the plants (based on the affinities of the fossil plant species with modern relatives), the pattern clicks in: Extinction seems to have been more pronounced in the lower than the higher latitudes, at least among plants.

All of which, of course, suggests that temperature itself might be the underlying culprit in the overall extinction pattern. Assuming that temperatures did drop and were implicated in the K–T extinctions, were the temperature changes gradually cumulative, or were they sudden, occasioned, for example, by the vast clouds of sun-occluding particulates thrown up by the impact (or volcanoes) in the paroxysm at the very end of the Cretaceous? And here we hit something of a stone wall, with some paleontologists sticking to the idea that temperatures had been declining regularly throughout the Upper Cretaceous and into the Paleocene. Others, like Upchurch, think there was a sharp drop in temperature, but one that was occasioned by the impact event itself, and most definitely not just a stopping place on an overall curve of temperature fall that had commenced long before the boundary.

This issue is by no means of academic interest only. It is actually of great moment to us: We really need to grapple with the relation between extinction and climate change, and in particular we need to know what the relative effects of different rates of climate change are on episodes of mass extinction. And here, at the very end of the Cretaceous, we still do not have a consensus on the state of the climate, especially global temperature. Everyone thinks there was a drop, and the drop is part of many bolide-impact scenarios. But it is also part of the gradual-extinction scenario.

There may as yet be no final consensus, but the events at the end of the Cretaceous are richly and complexly epitomized in the reams of data and partially conflicting interpretations that have appeared, almost all of it post-1980 Alvarez hypothesis. The bolide scenario (almost certainly partially correct at least, given the widespread iridium anomaly and presence of shocked quartz grains) has had a further impact: It has caused a tremendous amount of research to be done, on the K–T boundary especially but also on all the other large-scale mass extinction events themselves.

This research is terrific for science in general, and for all of us who would wish a clearer understanding of what went on and what were the underlying causes of these mass extinctions. And it helps us appreciate the magnitude of extinction. It helps us understand that extinctions are real. And, most of all, this research helps us grapple with the causal questions, so that we can tease apart the process from the underlying pattern. Also we can evaluate what the input of our own species might be, when added to processes that are already in place in nature.

The Cenozoic: Modern Life—and Extinction

Life, of course, did spring back, once again, after the K–T events. So far, it always has. And, as we have already seen, the pattern of rebuilding in the Paleocene was more like the brachiopod story at Nye Klov than Bramlette's original diagram of coccolithophore evolution: There was a distinct lag, not a totally instantaneous reconstruction of ecosystems. It takes time, of course, for new species to evolve, and ecosystems are composed of many species, most of which are newly evolved from the relatively few survivors that manage to make it through a mass extinction. What is surprising is not that there is a lag but that the lag times are so characteristically short: "Normal"-seeming marine communities were in place within a few million years after the K–T boundary.

Much has been written in recent years about mammals finally coming into their own, truly radiating into a wide variety of terrestrial (and aerial and aquatic, including marine) habitats only after the lock grip that dinosaurs and close kin had on these environments for the better part of the Mesozoic. It is not some imagined superiority of mammals versus reptiles but much more a matter of sheer chance that finally shifted the balance of control of terrestrial ecosystems from reptiles to mammals. Dinosaurs lost out, finally, not because they were archaic, primitive, stupid, or in any other sense inferior: They just didn't make it through, while some stocks of mammals, smaller and perhaps ecologically more generalized, did.

A less spectacular, but equally informative, example of the same phenomenon happened in marine environments as a direct reflection of the K–T boundary events. The ammonites had held a lock grip (almost relinquished twice, as they barely squeaked through the Permo–Triassic and Upper Triassic extinctions) on the externally shelled predaceous cephalopod niche from the Devonian through the end of the Cretaceous— a prodigious span of some 330 million years. True, we don't have a terribly good idea of when squids and octopi appeared, or how abundant they were. Their internal skeletons, when present at all, are so flimsy and biodegradable that they barely show up in the fossil record. We do know that there were true squids in the Upper Paleozoic, so swimming, predaceous cephalopods were not confined to the ammonites.

And then there were the nautiloids—like the ammonites, externally shelled, swimming cephalopods. In fact, ammonites were just a specialized offshoot, or subset, of the more generalized and primitive nautiloids. Nautiloids had arisen earlier (way back in the Cambrian) and had first radiated into a diverse array in the Lower Ordovician. They hung on throughout the Paleozoic and into the Mesozoic, though from the Devonian on they were always less numerous and diverse than the ammonites. Even if we had no nautiloid fossils from the Silurian on, we would know that they had survived, simply because there are five species of the living chambered nautilus still very much alive in the waters of the western Pacific. But there are always a few nautiloid species to be found at any point in geological time from the Upper Paleozoic through the Mesozoic. *Eutrephoceras* is a handsome, baseball-sized nautiloid fairly common in the Cretaceous Pierre shale of the Dakotas—the one kind of nautiloid in rocks that produces many species of ammonites.

It was the nautiloids, though, that survived the K–T events. After all those years of modest diversity, nautiloids staged something of a revival in the Eocene, some 10 million years after the K–T events. They diversified into a number of genera, and some even developed shell characters reminiscent of those of the more primitive, Paleozoic ammonites. One must be wary of speaking about competition between entire lineages: Ecological competition goes on between local members of the same species, and between local populations of different species. But the net effect of such competitive interactions might, in the long run, shape the diversity patterns of entire chunks of the genealogical tree of life: Just as dinosaur occupation of terrestrial habitats seems effectively to have precluded mammals from diversifying, so, too, does it appear that ammonite success had the effect of dampening nautiloid diversification.

Twice the ammonites appear to have beaten nautiloids to the punch, after the Permo–Triassic and late Triassic extinctions. It was only when the ammonites disappeared forever that the nautiloids were able to radiate. The meek really do, at least sometimes, inherit the earth, though squids really are far more abundant and diverse today than are their ancient nautiloid relatives: The nautiloid radiation was short-lived, not really surviving the Eocene.

Which brings us to one more major mass extinction event— one which effectively sets the stage for the later Ice Age (Pleistocene) events, which include, for the first time, ourselves—*Homo sapiens*. Though not rated in the top five global extinction events of the last 600 million years, there was a rather large-scale resetting of the biotal clock at the end of the Eocene. On land, many of the primitive mammals that arose in the Paleocene managed to live through most of the Eocene, finally giving way to modern groups that had begun to appear during the Eocene: ancestors of horses, rhinos, and tapirs (perissodactyls); of sheep, deer, and pigs (artiodactyls); of lions and tigers (carnivores). Whales and bats first appear in the Eocene, as do elephants. But the rich assortment of pantodonts, tillodonts, and condylarths—names totally unfamiliar to the ear attuned to modern mammalian life—never survived the Eocene.

Paleomammalogist Don Prothero has recently reviewed all the data pertaining to events at the Eocene–Oligocene boundary. The sedimentary record, thus the data on extinctions, are by far the best for the open-ocean planktonic microorganisms; the data are good, but not as good, for marine invertebrate communities, and are spottiest for terrestrial organisms. Deep ocean sediments rain down in a constant stream, accumulating on the seafloor and generally not disturbed until a ship comes along with a coring machine to take a deep and continuous sample. Shallow-water marine environments are subject to more stress, erosion, and discontinuity; and terrestrial habitats (even lake deposits) are notoriously discontinuous and subject to loss even after accumulation: Erosion quickly goes after rocks elevated above sea level.

The oceanic data reveal a series of extinctions, some more major than others. In a sequence of five separate events (including one right at the boundary), spanning a total of 10 million

years, there was a great net turnover of planktonic microorganisms. The data on shallow-water marine macroinvertebrates
are a bit difficult to correlate with the five episodes of micro-
plankton extinction, but Prothero reports a study by Thor
Hansen documenting very high rates of species-level extinction
in three separate events in the later Eocene along the coast of
the Gulf of Mexico. The first took out 89 percent of the snails
and 84 percent of the clams; the second saw 72 percent of the
snail species and 63 percent of the clams become extinct; the
third event, right at the Eocene–Ologigocene boundary,
claimed 97 percent of the snails and 89 percent of the bivalves.

Terrestrial plant data, according to Prothero, show a cooling
in the later Eocene; by late Eocene times, tropical floras seem to
have largely given way to more temperate forests and woodlands. Prothero himself found the major peak of mammalian
extinction to have occurred some 2 to 3 million years prior to
the actual Eocene–Oligocene boundary.

Prothero, not surprisingly, concludes that these data are
inconsistent with a simple, one-shot extraterrestrial event scenario. His own interpretation of the multistaged, 10 million-
year sequence of extinctions (and subsequent proliferations)
lies squarely in the realm of climate change: specifically cooling.
Geochemical evidence has indicated a decrease of more than
10° C (18° F) mean annual global temperature during this interval of geological time.

What caused that cooling and how that cooling worked to
produce as many as five discrete episodes of extinction become
the key issues. There is no doubt that, in our survey of major
extinction events of the past, temperature change, particularly
temperature drop, seems to emerge time after time as the
underlying culprit, a theme most persistently and completely
pursued by paleobiologist Steven M. Stanley during the 1980s.
We have seen enough, in fact, to turn to a detailed consideration of what, precisely, are the underlying causes of mass extinctions, and how these causal agents really work. We want to

know if there are overall causal themes linking most and maybe even all of the major global extinction events of the past.

As we turn now to a consideration of the causes of extinction, we should also realize that in leaving the survey of major extinction events as the Eocene gave rise to the Oligocene we have not quite finished a review of all the more prepossessing of such events: The Ice Age (Pleistocene) extinctions of the past 1.6 million years loom, in a sense, much more heavily than even the vastly greater extinctions of the very remote geological past. Perhaps as much as 96 percent of all species became extinct during the Permo–Triassic extinctions. That obviously had a tremendous effect not only on life as it had been but on life as it was to be forever after. But 10,000 years ago there were elephants, rhinos, lions, and sloths in both North America and Europe. To our *Homo sapiens* eyes, that simple faunal fact is more hypnotically compelling than the demise of the trilobites, the loss of the ammonites, even the extinction of the dinosaurs. The Ice Age extinctions are so near to us that we cannot even be sure that present events unfolding around us today are not just a continuation of events already well underway before *Homo sapiens* evolved some 100,000 years ago.

Most importantly, and the reason why we turn now to a discussion of what underlay the major global extinctions prior to the Pleistocene, is that we have, in effect, an experimental situation set up by nature that gives us a way to understand what the global impact of our own species, *Homo sapiens,* has been, is right now, and might be in the future. The Ice Ages, by their very name, suggest global cooling, already deeply implicated in the causes of prehuman global extinction episodes. But there is compelling evidence that people were also involved in aspects of Pleistocene extinctions, as we shall see after reviewing the causes of prehuman biotic turnovers.

CHAPTER SIX

Without a Helping Hand: Causes of Mass Extinctions in the Geological Past

IT IS ONE THING, of course, to acknowledge the events of history, especially the disasters. Like crime and traffic accidents, we all know disaster can strike. But as long as it is happening elsewhere, we take comfort that it won't happen to us. Yet here we are, in the midst of a rising tide of extinction, one that may or may not be part and parcel of a wave a few thousand years earlier: the Ice Age extinctions that have claimed so many larger mammals throughout the globe. It is not enough simply to admit that mass extinctions can and do happen: We need to know what causes them to better assess our own situation.

There is a strong tendency to blame disaster on the victims. This holds as much for crimes (consider rape, and even murder) as it does for accidents, and even for the incidence of some diseases—especially cancer, undoubtedly if enigmatically more feared than heart disease in contemporary society. But again, there is a certain logic to this assignment of blame: If it is the victim's fault, we need not fear a similar fate so strongly. It is especially the random factor of pure bad luck, with all the uncertainty attached, that scares us most. For then there is no particular reason to suppose the next victim will not be oneself.

People have always wondered about extinction: what causes

it and what the implications might be for future events. The interest has not been purely academic: The fate of the dinosaurs has been a matter of open public interest, even concern, since the early nineteenth century. The dominant and reassuring conclusion has generally been that the dinosaurs were lumbering, primitive, and by the end of their reign, even archaic creatures who finally simply had to give way to superior beings—mammals, eventually us. Never acknowledged in the assertion that time was simply up for dinosaurs (they having outlived their usefulness) is the parallel moral that mammals, including ourselves, may eventually also outlive our usefulness. If the rhynchosaurs of the Triassic (according to this line) fell to the dinosaurs, and dinosaurs in turn ceded their turf to the superior mammals, it is only being consistent to predict mammalian demise in the face of the rise of some new, as yet unevolved form of even more superior creature.

If we have tended to blame the dinosaurs (and by implication, all forms of life that are no longer members of the current biota) for their fate, such comfort we might have taken in our own supposed superiority is robbed through the very idea that evolution inevitably fashions creatures of superior design that, in the end, win out and supplant the world's former inhabitants. The very picture of evolution that allows us to blame the victims for their own demise contains the seeds of production of the very next set of victims. Blaming the victims is, if anything, even less realistic when it comes to mass extinction than it is when assessing why disastrous crime, accident, and disease strike some and not others.

Dave Jablonski's distinction between background extinction—normal, statistically rather regular rates of disappearance of species no matter what the cause—and mass extinction is a healthy step away from blaming the victim. Species extinction is as normal and regular a process as species originations; but, from time to time, extinctions come bunched up, compressed in such brief intervals, encompassing such a wide range of different species often over the entire globe and in every imaginable habitat, that something out of the ordinary is clearly happening.

If we assume that normal, background extinction is caused by a species no longer being able to find or recognize suitable habitat, it is possible still to pinpoint the cause to a failure of sorts: failure to accommodate to changed conditions. For example, when pigs and rats were released on the Indian Ocean island of Mauritius, dodos and other species that laid their eggs on the ground quickly disappeared. When humans started herding the flightless, penguinlike giant auks aboard the ships to serve as food supplies (and a commercial source of feathers), this species was also doomed.

Environmental circumstances occasionally change and collide utterly with the basic adaptations of species: Like sitting ducks, species are abruptly swept away. Sometimes the agent is environmental change, but often it is a change in biotic circumstances: collisions with other species that may compete with, prey upon, or even parasitize organisms previously unbothered by and unused to such interference. Of course, it is still possible to blame the victim here: if the dodos and great auks hadn't become flightless, they wouldn't have constituted such close relatives to sitting ducks. But things were working just fine for them until wholly undreamt-of predators came their way.

Oil spills or prolonged exposure at low tides offer familiar marine minianalogs to mass extinction, and help pinpoint the difference between background and mass extinction. Periodically, because of cycles involving the alignment of the sun, moon, and earth, low tides will be extraordinarily low and prolonged. Days on end may pass before the invertebrates rooted to rocks in the intertidal zone will be under water again. Crabs, snails, and starfish can escape to lower elevations and disappear beneath the waves. However, oysters, mussels, barnacles, and bryozoans must tough it out.

All such beasts, accustomed to periodic exposure, have ways and means of dealing with low tide. All need seawater in which simply to respire, let alone feed; the universal strategy to combat exposure, then, is to shut down, hole up, and wait for the tide to come back. Retaining moisture sealed up within their skeletal casings, survival is routine, except in the very rare periods of prolonged exposure. In these cases, there are big die-offs, cutting right across the genealogical board. Eventually, under extreme situations, everything clinging to a stretch of coastline may succumb.

No marine organisms are adapted to survive indefinite exposure to air, or being covered by a fouling morass of crude oil. They are adapted to a normal range of conditions that can be expected to occur during their lifetime. They are even set up to survive the occasional extremes, provided they have been met with in the past: Evolution cannot anticipate previously unencountered conditions. The legs of deer bones are many times stronger than necessary to absorb the shock when fleeing a pack of wolves, presumably a reflection of the extraordinary measures (like jumping off a cliff) sometimes taken to avoid being eaten. But there are limits. If you need seawater for oxygen, there is simply a limit to the length of time you can spend without it. In such circumstances, we go beyond the quality of an adaptation in general, and way beyond issues of variation within a population: the "who is better than whom" variation that natural selection works to modify, even improve, adaptations.

There is no blaming the victim in such disastrous situations: When a habitat is so severely altered that death approaches 100 percent of all organisms, in all species present, we must be dealing with causes outside the system itself. It is simply bad luck that it happened to the hapless creatures caught up in its effects. But is mass extinction just all bad luck? And, if so, should we be simple fatalists, hoping that it won't happen to us? In particular, if mass extinctions can happen even without our own contributions, does this give us reason for hope that our own apparent negative effect on the earth's present-day physical environment and biota really does not matter? We can probe these issues further by examining the impact hypothesis and contrasting it with various notions of earthbound causes of mass extinction.

Extraterrestrial Versus Earthbound Causes of Mass Extinction

It is more apparent than ever that we need to focus on the mass extinctions of the past to understand as clearly as possible what has caused them. There will be different implications for us in understanding and dealing with our own current environmental concerns, depending on what we decide here. There is a different potential message, for example, if we conclude that mass extinctions have a wide variety of causes, with no two exactly alike in their combination of contributing factors: That would make the search for understanding our most recent past and its relation to the present and very near future more difficult, but something we must realistically face.

On the other hand, it may be that there are one or two basic causes of mass extinction, and that makes the job of finding implications for the here and now considerably easier. As we have seen, the leading candidate in recent years for such a repetitious root cause has been collisions with objects from space: bolide impacts. If bolide impacts really turn out to be the culprit (and, recall, there is strong evidence of bolide impact at the Cretaceous–Tertiary boundary), we might conclude at once that the present has nothing whatever to do with the past: Whatever is causing the decline of song birds breeding in North America, it is obvious that it is not a collision with an asteroid.

On the other hand, there has been a fascinating resonance between the imagery of nuclear winter and the disaster scenarios—including the scorched earth from worldwide fires and occlusion of the sun, which drops temperature and impedes photosynthesis—that are attached to bolide-impact models of mass extinction. Prominent scientists have testified in the U.S. Congress on the perils of nuclear winter resulting from nuclear warfare: Citing the bolide-impact scenario at the K–T boundary, they warned that "the experiment has already been performed." We might hope to avoid serious collision in the near future, but we could well mimic the effects through our own agency. The point is well taken, even if the K–T scenario turns out not to be correct, or (as is more likely) not germane to all, or even *any*, of the other prominent episodes of mass extinction.

Then there are the implications of the answer to the question: Are mass extinctions periodic, that is, do they happen in some regular, rhythmic cycle, or are they totally random? If periodic, the implication for ourselves would fall directly out of the calculations predicting when the next collision will occur. We might even be tempted not to worry so much in advance, either trusting to luck for future generations who would have to go through the ordeal, or hoping that technology by that time could be mustered to divert the bolide and save the day (there are a number of sci-fi scenarios to that effect already). Most of us, I suppose, would not really care: Glaciers are figured to come down eventually and wipe out civilization in the northern regions of the hemisphere (the last glacial pulses came down as far as New York, forming Long Island; the next glacial advance actually seems to be 2,000 years late already!). If asteroids are as inevitable as glacial advances, why worry about them?

Some of the first calculations performed on the assumption that bolide-induced extinctions in fact are regular and periodic showed that we are about due for the next one! The figure, much to everyone's relief, was soon revised. The most widely accepted figure (accepted, that is, by people who think that there is any periodicity at all to mass extinctions) for the length of period between bolide-impact-induced extinctions is now put at 26 million years. Dave Raup and Jack Sepkoski, in their elaborate statistical analyses of Sepkoski's computerized summary data of the fossil record, refined the initial 32-million-year estimate of paleontologist Al Fischer. The due date further depends on calibration of the geological time scale, and the latest consensus view sees us not due for another blast for some 14 million years.

Clearly we need to know whether or not extinctions have indeed been regularly periodic. If so, that would pretty much have to mean that mass extinctions have underlying astronomical causes: Only astronomical cycles seemingly work on the scale that such periodicity would require. Again, a fascinating resonance between interpretation of empirical evidence and inference, much like the nuclear winter/K–T extinction scenarios, permeates the entire subject of possible astronomical causes underlying proposed periodicity of mass extinctions. If the K–T event is bolide-induced, and if there is a regular 26-million-year spacing between mass extinctions (implying extra-terrestrial causality), we might infer that all mass extinctions are bolide-induced, even in the absence of the "smoking gun"— iridium anomalies, shocked quartz, and the other forms of physical evidence supporting the bolide hypothesis mostly known (and only known unequivocally) from the K–T boundary sites. We must bear in mind that, despite initial claims (e.g., for the Permo–Triassic boundary in China), there are to date no verified occurrences of iridium anomalies associated with any of the mass extinctions, save, of course, the K–T event.

Astronomers have gotten into the act: Accepting the Raup and Sepkoski 26-million-year periodicity figure, the task is then to identify what would cause periodic bombardment by meteors or by comets, which quickly became the favored source of the bolides. Comets have wildly elliptical orbits. Periodically, comets race in from the extreme outer reaches of the solar system, accelerating as they approach the sun. They pass close to the sun, then wheel around and are flung back into the farthest parts of the solar system at tremendous speeds. Unlike the nine known planets, each of which is in its own orbit with no chance of collision with other planets, comets cut right across the planes of planetary orbit and have a real chance of colliding with one of the planets.

The Oort cloud is a hypothetical vast swarm of comets at the edge of the solar system. It has never been directly observed. Under this hypothesis, every so often a comet leaves the Oort cloud for its peregrination through the inner regions of the solar system. What, astronomers wondered, might dislodge many comets at once, increasing chances of impact (and, it must be noted, perhaps accounting for the stepwise pattern of extinction of species just prior to, right at, and just after the Cretaceous–Tertiary boundary)? Two candidates emerged: Nemesis, the brainchild of astrophysicist Richard Muller and colleagues, is the sun's hypothetical companion star. Stars often come in binary couples, rotating around one another; if sufficiently remote, astronomers may have overlooked such a star, whose gravitational effects may periodically be suspected of dislodging a swarm of comets and sending them toward our own sun. A rival hypothesis simply pointed the finger at Planet X (pun clearly intended), whose remoteness has so far precluded its detection, but whose existence might also periodically disturb the Oort cloud.

Astronomers predicted the existence of Pluto, the outermost of the nine known planets, before its actual visual discovery. Much as one can identify an unseen bird by its song, astronomical bodies are detected by their gravitational effects: Certain anomalies in the orbits of the outer planets meant that Pluto had to be there. For that matter, there is more mass in the universe than can be accounted for by the densities of visible stellar matter: Astronomers are convinced of the existence (though still very much mystified over the nature) of so-called dark matter. So astronomers are very much used to inferring the existence of structures based on the subtle hints proffered by their observations. But note here the chain of inference stems ultimately from the earthbound conclusion that mass extinctions have indeed been periodic.

We simply cannot be sure that such is the case. The debate rages on intensely, with Jack Sepkoski and Dave Raup (and a number of colleagues who accept their data and conclusions) sticking to their guns, and with any number of statisticians, geologists, and paleontologists taking sharp issue, to some extent with their data, and to a large extent with the nature of their very statistical analyses. As in much of the debate about extinction in general, it is very difficult to find consensus. As we shall momentarily see, this is as true of the debate over possible earthbound causes as it is of the debate between those who support each side of the more general outer-space versus terrestrial causes debate.

But messy and heated as the field of mass extinction causality is at the moment, we really have to do better than throw up our hands in despair of ever understanding what lies behind such events. The stakes are simply too high. Fascinating as the labyrinthine bolide debate has been, we cannot ignore the *Cooling* strong evidence that earthbound causes are the real culprits behind most mass extinctions. The clues keep pointing to climate change—especially global cooling—as the dominant cause of major extinction events.

However, there are a number of serious contenders for earth-based extinction culprits. It is quite possible that a number of factors work in concert; indeed, though there are still adherents of a volcanic source for the iridium at the K–T boundary, most geologists now seem to agree that there was one or possibly several bolide collision bouts near that boundary. Yet, even with bolide impact, we see global cooling invoked as part of the extinction scenario. Indeed, many of us have tended to see the K–T bolide impact(s) as an added, complicating factor that made a bad situation worse. Such a line of thought potentially has much to tell us when we come to consider the role that humans have played in Ice Age and contemporary extinctions.

But just as climate itself is not the only factor regulating diversity of the earth's modern biota, additional factors seem to be involved with episodes of mass ecosystem collapse in the past. Recall, for example, the role that simple area seems to play in regulating diversity. Size of available habitat is a crucial element in determining diversity patterns, and many paleontologists have argued that mass extinctions reflect simple loss of available habitat area. Given what we know about the effects of habitat loss on existing species today, change in an extent of available habitats must stay high on our list of potential causes of mass extinctions.

As we shall soon see (and as was obvious in the review of mass extinctions through the preceding three chapters), there is a great deal of empirical support implicating global cooling with mass extinction. But the reason why it appeals to me is that climate control invokes the very causal factors underlying mass extinction as we have already seen (in chapter 2) regulate diversity over the surface of the earth at any given moment. I am not suggesting that the biota, at any one moment, carries the seeds of its own destruction, or that it is a victim just ripe and waiting for an accident. Rather, the biota in general seems to be very like an invertebrate fauna clinging to the rocks of the intertidal zone in coastal Maine or California: fine, for the most part, from day to day, month to month, year to year. Biologists argue over the dynamics that regulate the number of organisms of different species represented. Competition and predation— biotic interactions among the various organisms, both within and between species—are variously specified as more or less important in such regulation. And, of course, there are fluctuations from day to day, and especially season to season and year to year, in what is there and what is not, what persists and what disappears. But then there are those unanticipated low, low tides, or those oil spills, and the rules of the game are changed.

That, I think, is what happens when the normal controls of global diversity are exceeded beyond a certain tolerable threshold limit. The normal factors of diversity regulation (which, as we saw in chapter 2, depend so heavily on adaptations to temperature, and especially yearly temperature variations) are thrown out, or superseded under certain conditions. And mass extinctions are the result.

Earth-based Causes of Mass Extinction

Science often seems like a perpetual motion machine, not one that simply rocks back and forth but one that has a definite linear element giving us an ever-more accurate picture of the universe as time goes by. Science is supposed to be different from, say, the study of literature, where every generation simply must translate anew Homer's *Odyssey* into the language of the moment, exploring afresh the fabric of that great epic poem and not depending on the renderings of bygone generations. Science supposedly builds directly on the work of past generations, keeping what still appears to suffice as a clear picture of the nature of the world, and devising improved theories and explanation to replace faulty views that are simply discarded.

Yet there are styles of scientific thought and eternal questions in science that forever must be rethought by each generation. The favored culprit for mass extinctions has always been climate change, especially global cooling. Yet the recent reappearance of this line of thinking as perhaps the dominant, most highly favored candidate reflects no simple historical conservatism and continuity: We are not yet out of a true maelstrom of empirical and theoretical research on mass extinctions, one that has seen a wide variety of causes, earthly and extraterrestrial, debated with extreme passion. It is because of new observations and theoretical explorations, tying together what we think we know about the factors controlling diversity on the earth at any one moment with physical and climatic events associated with mass extinctions, that the global cooling hypothesis has once again begun to attract support. Science is a mixture of progress-in-knowledge with constant reevaluation of eternal questions. In a sense, science too is the continued translation of ancient texts into modern garb.

There have been, of course, many theories on the causes of mass extinctions proposed in the past that simply cannot be right. In our quest to understand what really did happen, and in our desire to apply accurate principles to our own modern dilemma, we have little time to spare on the falsified notions of the past. Yet older ideas on mass extinction often reflect a fundamental worldview on the nature of the organization of living things and their relation to the physical world. Understanding the basic nature of these approaches (at the moment wholly out of favor in biology, but who can say what future biologists will think?) clarifies what we ourselves think about life, its organization, what keeps it going, and what leads to its occasional teetering on the brink of total collapse.

Easily the most arresting category of out-of-fashion extinction causes places the blame entirely on the victim. Best known is racial senescence, where species become extinct because their evolution takes them to extremes that make them unable to continue living. Most egregious, I think, is the case of the Mesozoic coiled oyster *Gryphaea*. Oysters are a branch of bivalve mollusks whose lower shell is enlarged, and is either attached to an object or rests directly on the seafloor. The upper portion of the shell is a lid. In modern oysters, such as the delectable *Crassostrea virginica*, the bottom valve is barely larger than the upper. But in *Gryphaea* and some other Mesozoic oysters, the lower valve is elaborately coiled and ponderous. The upper is a tiny little lid resting on top. Early paleontologists supposed that some populations of *Gryphaea* in the Jurassic of England were so tightly curved that the upper lid simply could not open. Later x-ray analysis (as reported by Stephen Jay Gould in the 1970s) amply demonstrated what seems in retrospect only to be expected: Mud had filled in the space between the lid and the coiled portion of the lower valve; it was mud, not shell, that gave the impression that the lid could no longer function.

There are other such stories: Irish elk (actually huge, now extinct, Pleistocene relatives of the European red deer) supposedly were the agents of their own destruction, the antlers of the males allegedly growing so huge that they became difficult to carry and were caught in trees as the deer tried to walk through the woods. Or take the ammonites, relevant here especially because their final demise was wrapped up with the K–T mass extinctions. One ammonite group, the scaphites (who we now know if anything were expanding and certainly prospering as the K–T boundary drew nigh), are a bit oddly coiled by normal ammonite standards. Baby scaphites coil in the normal manner (something like a pearly nautilus), but the mature adult lives in a terminal, J-shaped chamber. As if this weren't enough, other ammonites either give up coiling altogether (the aptly named *Bacculites*), or become truly bizarrely coiled, twisted every way so that in genera like *Nipponites* no two specimens are exactly alike. Here is evidence of creatures evolving past the point of no return, or, more exactly, running out of evolutionary gas.

Thus racial senescence contains the idea that species, like organisms, have youth, middle, and finally old age. Species, even entire lineages, were thought to go through a normal life cycle, with the inevitable degeneration toward the end, manifested by phylogenetically bizarre forms. One after another, all the old classic examples have been debunked: Natural selection, after all, could not fashion an anatomical design that would sow the seeds of an organism's destruction in it. Perhaps, it was argued, in the case of the coiled oysters, the enforced closure that would kill the organism only occurred after the period of reproduction: Natural selection would be powerless to hinder such pure effects of extended growth if they always occurred after reproduction had already ceased for other reasons. But even here, it turned out to be mud, not the oyster's own shell, that kept the lid clamped down—and that occurred only after the oyster was already dead and buried from natural causes.

All the old theories about extinction for reasons that amount to the organisms' own fault fit right in with a counterpart set of ideas: evolution for intrinsic reasons. There is a slew of evolutionary theories as antiquated and scientifically dead as racial senescence. Nomogenesis, aristogenesis, entelechy—these and others were the pet ideas of prominent biologists (and generals—Jan Christian Smuts, of considerable sociopolitical historical importance in South Africa, gave us holism, part and parcel the same sort of thing as all the others). Closely tied in with the idea that evolution begets, or rather represents, improvement or progress, these theories all saw change through time as originating and essentially controlled through factors internal to organisms themselves. There was widely believed to be something internal within organisms—something inherent in the gene plasm—that would automatically make organisms bigger (often) and better than their forebears.

Darwin's view, of course, saw the match between organism and environment as the response of populations of organisms to environmental stimuli: If the environment changed, perhaps some variation in a population would be selected because it had a heightened survival value under the new environmental regime. And even when environments don't change, it is still possible to imagine natural selection fashioning a better mousetrap: Variants in a population might be even better swimmers, or trappers of sunlight, or whatever, than had ever been seen before. And the qualities that conferred their heightened success would be passed along, and there would be improvement.

Progress, or improvement, lies at the heart of both Darwin's original vision and the competing notions of vitalism. The difference lies in the relative importance attributed to external versus internal factors. Darwin's notion, in both original and modern forms, indeed does see the origin of variation as intrinsic to organisms. We now know that the ultimate source of variation is mutation, and that (though it may spring from external mutagens) mutation is pretty much a matter of inherent biochemical mistakes arising in reproductive cells. Mutation throws up a shotgun splay of variation, and the environment is the arbiter, via the differential success of the organisms carrying this variant or that, out in the world, trying to make their livings and reproduce. In the older vitalistic theories, though, it is more a matter of the inherent superiority of the genes themselves that win out in the end: The weak fall by the wayside, true, but the emphasis is on the internal genesis of superiority, something widely held to be inevitable through the inner workings of the genes themselves.

In terms of our own modern biological cosmology, we live in an age where the external environment has at least equal partnership status with organisms. Who has the upper hand— the environment or the organism? We tend these days to follow Darwin's basic drift: Organisms are subtly molded in the face of environmental contingency. The only real competitor for natural selection in Darwin's time was the inheritance of acquired characteristics (usually, if unfairly, associated with the great French biologist Jean Baptiste Chevalier de Lamarck); here, if anything, the environment had an even greater upper hand than it does under natural selection: The environment induces direct effects on organisms, and those effects (calluses, long necks in giraffes, etc.) are inherited directly by the affected organism's offspring. The advent of a modern theory of inheritance around the turn of the century put an end to this line of thought (though there are occasional flare-ups as time goes by). But our objections to it surely cannot rest on any major difference that inheritance of acquired characters might have with natural selection on the issue of the relative importance of external and internal factors in evolution, and on its converse, extinction.

Life, of course, has had profound effects on the physical world. Easily the most profound is the chemical effects of respiration: Photosynthesizers release oxygen. The botanical garden in Naples is said to have air several times purer than levels measured on adjacent streets. Globally, it is photosynthesis along with volcanic outgassing that converted the primordial, reducing (and, from a multicellular-organism-centered point of view, poisonous) atmosphere to one with sufficient oxygen to support the existence of heterotrophic life.

On a less dramatic scale, but still telling, is the evolution (sometime in the Mid-Mesozoic) of calcareous-shelled plankton—microscopic photosynthesizing algae and shelled amoebae (forams) floating in the upper water column of the world's oceans. As we have seen, the coccoliths and forams of the Cretaceous, which figured so heavily in the event at the K–T boundary, had only appeared sometime in the preceding Jurassic Period.

Under extreme depths (meaning pressures and temperatures) in the oceans, calcium carbonate spontaneously dissolves. Above that compensation depth, however, it can accumulate, as the vast oceanic calcareous oozes and their now-exposed chalky counterparts amply demonstrate. Invention of the calcareous plankton changed the carbonate cycle of the entire earth. There is no way that the carbonate now lying in such huge quantities on the ocean floors will be recycled until it is sucked down into the marginal oceanic trenches, incorporated into the crust, and eventually, at least in part, exuded through volcanism and other tectonic activity. Life, indeed, can have a tremendous effect on the physical conformation of the earth, which in turn has implications for the life of organisms now extant. I once saw a modern land snail crawling over the carbonate shell of a 500-million-year-old snail embedded in a limestone outcrop—the source, of course, of the carbonate of the living snail's shell.

We are, in short, living in an intellectual climate where the environment seems overwhelmingly and decisively crucial in shaping the moment-by-moment lives of creatures. The environment shapes, as well, their long-term evolution, and their ultimate extinctions. We hold these views tenaciously, even though we ourselves, species *Homo sapiens,* have become progressively more isolated, even insulated, from the environment throughout the course of human history, and especially the rise of Western civilization. We know of nothing intrinsic to species that would suggest they can in fact run out of evolutionary gas.

There is for species nothing like the natural process of aging in organisms. Organisms age because carbon-based chemical systems are extremely vulnerable. Natural selection cannot eliminate death, but it can toy with the timing of aging and death, juggling it with the reproductive cycle. Had reproduction not been invented, the living systems that might have evolved would long since have become extinct. And even with organisms, it is often the physical or biotic environment—accident, predation, disease—that brings about death before whatever might be the appropriate equivalent of threescore years and ten has run its course.

From time to time, evolutionary biologists bravely assert that organisms are, in fact, not as much as hapless and helpless in the face of the implacable physical environment as we tend to depict them. Yet all such objections have yet to transcend the "yes, but . . ." phase: Organisms do build reefs, nests, dams, and burrows, but their impact on their physical surroundings are still paltry when compared to the effects that the physical world (and other organisms) have on them. We should acknowledge, though, how very tough individual organisms often can be. And oil spills tell us how resilient genetic systems—entire species—can be in resisting extinction. But when all is said and done, it is the style of our generation (and generations dating back at least to Charles Darwin, for that matter, despite the vitalist aberrations of the earlier part of the twentieth century) to see organisms caught in the thrall of their environment.

The environment, though, includes other organisms as well as the inorganic elements that all organisms face. Most contemporary humans, no doubt as a simple reflection of the complexities of Western industrialized existence, tend to think only of other humans and maybe a few pets. But ethnographies amply demonstrate the continued existence of smaller entourages of humans keenly attuned to the animal and plant life surrounding them, and on which they depend, or must avoid. And significant numbers of ecologists focus almost exclusively on the myriad, complex interactions between organisms—members of the same species, and especially members of different species living cheek by jowl in local communities. We often, in other words, tend to think more of other organisms, whether of our own species or not, when we think of "the environment."

Entire theories of the structure and functioning of ecological communities, in fact, are couched almost wholly in terms of organismic interactions. Communities are highly accommodated if their very structure depends on the intimate relations between members of various different species. Everyone has seen the curves of rabbit and fox abundance as they cycle up and down: Boom in a prey item (rabbits) causes a delayed boom in the predator (fox) population, which quickly depresses the rabbit population down to levels so low that the foxes soon face starvation. Thus the rabbit population expands. And on and on it goes. Ecology is full of such equilibrium concepts, though ecologists have recently begun to concede that life is not so orderly and evenly regulated within communities and ecosystems. So we must ask, is there any evidence that the very factors of biotic interaction that appear to hold communities together also can lead to the sudden crash of the system?

The answer appears to be "no." Ecosystem collapse, the very essence of the extinction process, may well hinge in some instances on keystone species—species, for example, at the base of the energy (food) chain. But there is no evidence that ongoing interactions between species ever get sufficiently out of whack or taken to extremes that exceed some tolerable limit such that component species are driven to extinction—that is, until *Homo sapiens* arrived on the scene.

Sometimes foreign species do invade new habitat and even eliminate some hapless longtime denizens. This has recently happened as the brown tree snake, *Boiga irregularis,* has invaded Guam. It has already killed off the Guam rail and some other endemic bird species. But the effects of such invasions are invariably local. Ecological interactions may conceivably be implicated in background extinction events, but there is simply no evidence that they could ever account for the collapse of the entire system. There appear to be no seeds of massive self-destruction arising from biological interactions themselves. We must seek further for environmental causes. And these appear to lie squarely in the domain of the physical environment.

Yet there is one element of species–species interactions that deserves closer scrutiny here, as it gives us an important key to understanding some of the very ground rules in species extinction. Rats and pigs (brought in, of course, by humans) eliminated dodos. Passenger pigeons succumbed to human hunting, and brown tree snakes (presumably also inadvertently brought by humans to Guam from their native islands) have eliminated a number of vulnerable bird species. Introduced species can, and do, regularly cause extinctions. But in all such extinctions, the species that succumb are very narrowly distributed. (Passenger pigeons were more widely distributed, but extremely vulnerable to hunting by firearms.) And that tells us something about species—what they are, what they do—and how species biology figures into the extinction process itself.

Oil spills, even if they are from human hands, at first seem to change the emphasis from biological to physical environmental cause. But oil spills have a lot to reveal about the way biological systems are organized, and that tells us a lot about what to expect when such systems are threatened, whatever the nature of the cause.

Most examples of biologically induced extinction are, as we have seen, rather narrowly distributed. Most examples, in fact, come from isolated islands. Most species have far wider distributions than the Guam rail or the Mauritian dodo. And here is the critical point: Local ecosystems are composed of organisms belonging to many different species. There is a strong but misleading tendency to say that an ecosystem is composed of species. But this is simply not the case. Ecosystems are composed of local populations of many different species.

Nearly invariably, any given species is distributed hither and yon, and its organisms are parts of many different ecosystems, often in rather different habitats. When those low tides, or oil spills, clobber the shores of northern California, so far at least, there have always been populations of the very species affected locally who remain alive and well elsewhere. It is these outlying populations that contribute mobile, floating larvae that quickly come in and establish ecosystems very similar (though never identical) to those eradicated in the disaster.

Species are banks of genetic information. Their only function, really, is to dispense genetic information. Species provide the raw recruits that are constantly restocking local ecosystems. Nowhere is this more evident than after some local disaster knocks an ecosystem for a loop, degrading the habitat and causing the deaths of most, if not all, organisms living there. Rebuilding seldom comes from within. It comes, instead, from outside.

Note, though, that because species are rarely ever confined to a single ecosystem, species are buffered against extinction. It takes a much larger, truly regional disaster, encompassing the entire geographic spread of a given species, before the species itself faces extinction. Ecologists such as Paul Ehrlich are of course right when they point to the loss of local populations (and races and subspecies—subdivisions of species) as also important, as in all such cases genetic diversity is being lost. But it is the species level that is nonetheless critical: As long as some pockets of a species survive, there is a chance that local semblance of an ecosystem can be reassembled. For example, large-scale forest fires started by lightning strikes (and thus purely natural) force the reconstruction of woodland and prairie ecosystems periodically. Species are the critical line of demarcation: By their very natures and distributions, species are resistant to extinction. Cross over that line, and you have a threshold effect: New ecosystems must be fashioned from different materials.

As we have seen in chapter 2, organisms in some species are more widely environmentally tolerant (broad-niched, or eurytopic) than others; and width of niche is closely related to width of geographic distribution. The plot thickens: It becomes easier to see why narrow-niched, geographically restricted species are more vulnerable to extinction. As we have already seen, there are far more species in the tropics than in the higher latitudes, because the tropics present an equable climate, hence no need for species to be climatalogically broad-niched. The tropics, on land and in the seas, are packed with narrowly distributed, narrowly niched species. The tropics appear always to have led the higher latitudes in extinction, both in absolute and in relative terms. And the ancient extinctions help us understand why this is so.

So it can't be that biotic interactions themselves will ever lead to anything more than isolated cases of background extinction. We are left with one choice: Mass extinctions—true ecosystem collapses, with far-reaching effects on the fates of large-scale taxonomic entities—must reflect threshold effects. Organisms, like our intertidal beasts clinging to the rocks as the tide goes out, can only be engineered to withstand so much fluctuation in physical environmental conditions. Some are more narrowly adapted than others; it simply must be that mass extinctions reflect circumstances where limits on a broad scale, affecting all sorts of differently adapted species in a wide variety of different ecosystems, are passed. Systems can adapt, staving off losses, but can only take so much before they snap. It is the possible physical and earthly causes of such threshold-transcending snaps that we must examine to narrow down our search for the causes of mass extinctions in the geological past.

*Glaciers and Tectonics, Sea Level and
Temperature Change: The Causal Dynamics of the
Great Mass Extinctions*

Steve Stanley, in midcareer as one of the more thoughtful and
innovative of the new breed of paleobiologists, has contributed
mightily to the revived field of macroevolution—essentially,
the large-scale elements of evolutionary history and process. It
was he who coined the expression species selection in the 1970s;
his compilations of evolutionary rates have been instrumental
in sharpening our apprehensions of patterns of evolutionary
change. Like Dave Jablonski (and, of course, Jack Sepkoski
with his monster data base), Stanley is a master of generating
his own massive amounts of data, and of incorporating the
studies of others to come to far-reaching conclusions about the
nature of the evolutionary process. Both Stanley and Jablonski
are experts on Mesozoic and Cenozoic mollusks. And both
turned to serious contemplation of mass extinctions in the
1980s.

Dave Jablonski, recall, is the one who specified the difference
between mass and background extinction, maintaining that
mass extinctions are not merely periods of intensified normal
extinction but times when a threshold has been crossed. His is
also the notion of Lazarus taxa that provides an effective measure of how complete the data are around an extinction boundary. As Jablonski has said, that Roger Batten found more
different species of Paleozoic snails in the Middle Triassic than
in the Upper Permian has a lot to say about how incomplete the
fossil record can be. Extinctions might seem more severe than
they really were if it turns out some taxa missing in action turn
up again alive and well on the other side of that line.

Jablonski has adopted a reasonably pluralistic view of mass
extinctions. He acknowledges the evidence for extraterrestrial
incursion and even periodicity, but he also realizes that the
geographic and climatic factors invoked by others have much

going for them. Steve Stanley has, in contrast, taken a much more passionate stand in favor of one particular explanatory theme for mass extinction: climatic change, specifically global temperature drop. It is an old theme in the mass extinction genre. But it must be acknowledged that the signal is very strong, and Stanley has gone a long way toward strengthening that signal and figuring out how the mechanism actually works.

Deciphering the earth's past depends on inference. There are no ancient thermometers buried in rocks, their readings stuck at the ambient temperature at their time of burial. But there are clues, some direct, some far more inferential, that yield at least a guide that one interval of time was cooler than another. And, for relatively recent geological history, we can even do better than that: Isotopes of oxygen vary in proportion of abundance depending on the ambient temperature, and the surface temperatures of the world's oceans going back at least a few million years can be read with a fair measure of confidence by analyzing the ratios of oxygen isotopes in the shells of planktonic foraminifera in the deep-sea fossil record.

Organisms tell us something about temperatures, too: Lions and hippos in what is now Trafalgar tells us, with little doubt, that British weather was warmer 120,000 years ago during an interglacial than it is right now. As Steve Stanley has pointed out, the spread of glass sponges in the Upper Devonian seas betokens a drop in temperatures—at least, if what we know about the cold-loving propensities of modern glass sponges is a reliable guide. But we must be careful: Presence in the past of tropical types of organisms in higher latitudes can mean two things. Either the earth was warmer near the poles back then, or the ceaseless movement of plates (continental drift) has secondarily shifted their fossil remains into climes in which they would only be caught dead.

Therein we confront a major squabble among two basic camps of paleontologists who seek earthbound causes for the mass extinctions of the past. If Steve Stanley has been the most vocal proponent of global cooling as the culprit, then Anthony

Hallam, respected paleontologist from Birmingham, England, stands at the head of the ranks of those who prefer another time-honored explanation: Hallam is convinced that the evidence for extinction being caused by global cooling is forced and inconsistent. To Hallam and like-minded paleontologists (stretching back to the previous century), the real culprit is loss of habitat area. In particular, in the marine environment (where mass extinctions have in fact been studied most intensively), lowering of sea level is the real culprit, especially loss, or regression, of shallow-water marine habitats from the interior regions of the continents.

Stanley has reviewed the arguments for habitat area loss as the culprit, and has, in turn, found them "trivial." We will review both separately (and dispassionately!). But it is important to acknowledge the competing hypotheses at this juncture simply because, to a confusingly great extent, they both rely on the very same data: The best evidence for temperature drop has always been sea-level drop!

Now, there are two major agents of sea-level change. One is the growth of glaciers, especially at the polar caps. As we have already seen, the current status of the continents as relatively high and dry is exceptional when compared with the flooding of continental interiors—the rule, rather than the exception, over the last 600 million years (and presumably earlier as well). Even when water still remains on the continents, clever geologists have devised ways of charting shifting depths from the nature of the sediments (and the fossilized faunas in them). Sea level rarely remains at a complete standstill. We are in an interglacial period right now, and the popular press has been charting sea-level rise and the threat to coastal structures for years. With the specter of global warming (and recent claims that the Antarctic ice shelf is shrinking faster than expected), the rate of sea-level rise is more than likely going to increase in the near future, until the next glacial pulse, whenever that might be, reverses the trend.

But there is a competitor for strict glacial control of sea level.

The fundamental mechanisms responsible for plate movements in the earth's crust also have their effects on sea level. The earth's crust is literally cracked and divided into six major and a number of smaller separate chunks—the plates. (*Plate tectonics* is the general term that has replaced continental drift for the very good reason that it is not continents alone but all aspects of the earth's surface, certainly including the deep ocean floors, that are involved in these relative motions.)

The source for all this motion is the heat welling up from the deep interior of the earth—heat derived from radioactive decay. In time, the decay will cease, and with it the turmoil, all this motion, on the surface—motion that gives us the high mountain ranges and the deep oceanic troughs. Motion has already ceased on the moon. Just like a bowl of hot soup (Japanese miso soup shows this especially well), there are areas of upwelling, where internal heat comes to the surface, by convection. At the surface, the liquid cools and spreads away, cooling all the while, and eventually sinking in a downward convection cell. (Cooler masses of the same substance are denser than hotter regions, and will sink.) In addition, the surface area of a bowl of soup is fixed, so spreading outward from an upward convection cell must be compensated for by sinking in downward convection cells.

The earth is very much like that hot bowl of miso soup. There are areas of upwelling of hot mantle material and these are nearly all confined to the centers of deep ocean basins. (The East African rift valley, the Dead Sea, and, some say, Lake Baikal in Siberia are areas now on continents where the process is also happening. Taken to its limits, these will be new sites of continental sundering and the appearance, in due course, of true oceanic floor. The Serengeti, whatever its fate in the hands of humans, is destined ultimately to host a benthic marine community!) The Mid-Atlantic ridge is the most spectacular example: New rock is continually being formed from lavas flowing along the very top of the ridge. As the lava cools, it spreads to either side and sinks. Europe and North America, and Africa and South America, continue to spread apart at the rate of around 1 centimeter per year.

Unlike the bowl of soup, the earth is not contained in a rigid shell; potentially the earth could be expanding as a result of all this convective activity constantly adding new crustal material. But this is not so: There are regions of downward convection as well, where crust is being consumed at the same average rate as it is being generated elsewhere. These are the deep oceanic trenches, formed where dense oceanic crustal material is sliding down (in those series of jolts we call earthquakes).

We can measure the rate of spreading fairly accurately these days. Lasers in Iceland (which sits astride the Mid-Atlantic ridge) measure that 1 centimeter creep per year rate at which east and west Iceland are now diverging. We can measure how many miles (some 350) the rocks of San Francisco have been displaced northward along the west side of the San Andreas fault and get an average rate of 0.2 inch per year during the last 120 million years that the fault seems to have been active. But average rates smooth out the actual differences in rates that might be lurking below the surface. In particular, seafloor spreading seems to have been much more intense during certain geological epochs than others. And this will have direct effects on relative sea-level heights.

Heightened times of seafloor spreading should produce greater volumes of midoceanic ridge. As these enormous undersea mountain ranges grow, they will displace huge volumes of seawater. So much of the variation in sea level, including the waxings and wanings of shallow seaways over the continental interiors, is attributable to variations in rates of plate tectonic activity.

How to tell glacial from tectonic control of sea level can be a real problem. It is generally acknowledged that glaciers can tie up or release an equivalent amount of water a lot more quickly than can variations in rates of seafloor spreading. We use the term *glacial* to mean "agonizingly slow," but moving plates are the tortoises, and glaciers the hares, when it comes to displacing volumes of seawater. Over geological time, though, both mechanisms have ample opportunity to wreak their effects.

So, not every drop in sea level need betoken a glacial advance. Proponents of the species-area effect—loss of habitat— as the vera causa of mass extinction are quick to point this out. Yet there are many times when a pronounced, prolonged drop in sea level is correlated with both mass extinction and with incontrovertible evidence of glaciation.

Glaciers leave ample evidence of their existence. Long Island is simply two combined terminal moraines, a huge pile of rock and sand dumped from the fronts of the two southernmost of the four glacial advances of the last Ice Age. Tillites are hardened sediments left by glaciers. Glaciers scour the landscape, making U-shaped valleys and leaving deep, unmistakable parallel scratches on areas of exposed bedrock. The more recent the glaciation, the more the telltale signs left behind. In the Paleozoic, the evidence is harder to come by, but there is evidence. In the Upper Ordovician, for example, glacial scratches and tills amply demonstrate glaciation in (of all places) North Africa. (Remember that plates do move around; paleomagnetic evidence has since shown North Africa to be lying squarely astride the South Pole in Upper Ordovician times.)

Tying all the mass extinctions of the past, however, to any single cause can be very tricky. The Ordovician glaciation is perhaps the best documented, and few geologists and paleontologists dispute the connection between sea-level drop and glaciation at that time. On the other hand, there is no equivalent, direct evidence for glaciation in the Upper Devonian. The standard story (at least until very recently) on the Upper Permian is that the climate was ameliorating: The great glacial fields of the earlier Carboniferous were melting. Yet sea level, all agree, continued to fall; Stanley points to recently gathered evidence of glaciation in a northern crustal block now forming a part of northern Siberia. Moreover, polar ice caps need not be associated strictly with continents. (They are, today, in Antarctica, but not at the North Pole, which is covered by the frozen-over Arctic Ocean. Lack of evidence of continental glaciation does not mean there were no fields of ice, though sea ice by no means is as thick as true continental glacial fields.)

Critics (like Tony Hallam) of the global cooling hypothesis as a general mechanism underlying most mass extinctions are fond of pointing to times of extensive glaciation when mass extinction did not seem to follow at all. (Critics, like Steve Stanley, of the species-area or habitat loss theory likewise point to times of lowered sea level with no apparent concomitant extinctions following.) But glaciation is a reflection of global temperature drop, not really its cause: Global cooling can also arise from changes in oceanic circulation patterns, bringing polar waters into more tropical areas. Such circulation patterns can derive from changes in plate position.

Glaciation and nonglaciation, sea-level change caused by glaciation or by tectonic factors, climatic change caused by glaciation or by tectonic factors—these are the confusing sets of variables that are being debated. As a way of cutting through the confusion, Stanley has looked afresh at aspects of the mass extinctions themselves, concluding that there is independent evidence within patterns of mass extinctions themselves that the cause of many of them has indeed been global cooling (and not, he steadfastly maintains, simple loss of habitat area).

Foremost among these strong signals of the effect of climate change is that tropical species are invariably more greatly affected than higher-latitude biotas (always correcting for changes in the geographic position of the plates since the faunas were actually living!). Reef communities have always been restricted to a belt fringing the equator, and not running any higher than 40° north and south latitude. Reefs are prima facie evidence of tropical climes, and reef communities are invariably severely affected in the major mass extinctions.

Further evidence of global cooling is the protracted periods of time it takes for reefs to come back once they are seriously reduced in a wave of mass extinction. Also, cold-water forms tend to replace warm-adapted biotas (indicating spread of cold conditions), or perhaps simply the asymmetrical relation between high-latitude flexibility and the narrow comfort range of tropical species. Tropical species, recall, are intolerant of climatic (or other forms of environmental) fluctuation simply because for millennia, even millions of years, conditions remain stable. Tropical species specialize and as a result they become very narrowly restricted in their geographic ranges. Higher-latitude species, whose organisms are adjusted to a yearly swing in climatic and other environmental conditions, are far more likely to be able to adjust to warmer conditions than tropical organisms could be expected to take to colder conditions. But, in any case, global cooling clearly would reduce the width of the tropics, expanding the temperate and arctic regions. And cold-adapted organisms are bound to do better than tropical species.

Stanley also raises the matter of timing: Mass extinction events are by no means overnight affairs, despite some of the earlier, more extreme scenarios associated with the K–T boundary impact hypothesis. As we have seen, extinctions are developed in waves. They are not smoothly gradual but often come in pulses. There are often two or even more such pulses, and the whole event may take as long as 5 million years. Though champions of the impact hypothesis have explained the step-wise pattern of extinction at the K–T boundary as the result of a series of bolide impacts, we must ask if such effects are to be expected from either of the two prominent earthbound suspects of mass extinction causality.

Stanley claims that global cooling would not act instantaneously, but rather would affect different regions and organisms at different times. His best-worked-out example actually involves the Pleistocene (the latest Ice Age), which gets us a bit ahead of our narrative. In brief, Stanley has made a strong case for differential effects of global cooling by comparing what happened to the mollusks on either side of the North American continent as the glaciers were growing during the last several million years. Losses were far greater in the Atlantic than in the

Pacific—a disparity that Stanley attributes directly to the differential ability of organisms on the Pacific side to retreat to more southerly, warmer climes while those along the eastern seaboard were trapped in the Caribbean and could not escape farther south.

Similarly, on a global scale, the Mediterranean and the North Sea were also cul-de-sacs during the Pleistocene while faunas of the western Pacific, like those fringing North America in the eastern Pacific, were free simply to migrate up and down the coastline, as temperatures were variably increasing or decreasing. Indeed, the cliffs at Capitola, California, are chockful of 3-million-year-old fossils, and most belong to species that can still be found in the waters of Monterey Bay lapping up against those very cliffs.

The hypothesis of global cooling has another strong point going for it: Temperature change, though it affects land and sea and different regions at different rates, is nonetheless literally global in extent. Terrestrial tropical biotas will face precisely the same sort of problems as their confreres below the waves: nowhere to hide as their stable habitats yield to more temperate, seasonable, variable conditions. Because tropical species are typically much more narrowly distributed than higher-latitude species (Rappoport's rule), they are automatically more vulnerable to environmental change.

Paleontologist Elisabeth Vrba (whose work on antelope species figured into our discussion of width of ecological niches in chapter 2) has added an additional ingredient to one puzzling aspect of the relation between climatic change and extinction: Why, if temperatures are dropping gradually, are there certain snaps or pulses to extinctions? Events in Africa involving antelope and other taxa (even primitive hominids—members of our own evolutionary lineage) are the proper stuff of the following chapter, but in brief, in her turnover pulse hypothesis Vrba argues that the biota is tremendously conservative. Physical environmental change of all sorts, but certainly including temperature change, will be resisted by organisms as long as possible.

As in Stanley's scenario, organisms in various species will migrate, ever searching for familiar habitat, as the first-order response to environmental change. Some will be able to take the changes better than others (high-latitude vs. low-latitude tropical species, for example, in the face of temperature decline). Even rooted organisms (trees on land, sea lilies, brachiopods, and many others in the sea) will migrate, trying to track suitable habitat in the face of environmental change simply by seed or larval dispersal. There will be, as Stanley points out, differential success based on the very facts of geography.

But the biota of any one area and set of conditions will have a certain degree of tolerance to change and thus resistance to extinction. Within a local biota, too, some species will be more tolerant (eurytopic) than others. But, eventually, tolerances are overcome. Especially if those tolerance limits affect species crucial to the food chain, sudden waves of extinction can sweep through a biome as a delayed threshold effect, sometime after the onset of the negative environmental change that ultimately leads to serious episodes of extinction.

Thus Stanley and Vrba point the way for understanding how essentially gradual environmental change can lead to delayed, sudden snaps—quick waves of mass extinction. There can be more than a single wave, too, as thresholds are passed for differently adapted biotas (i.e., the tropics would be hit earlier than the cold-adapted biotas, all other things being equal), and as ecosystems in different regions fare somewhat differently from their counterparts elsewhere in the world. Protracted cooling might very well have caught up eventually with the Ice Age Pacific invertebrate faunas, producing extinction statistics very much as severe as seen in Atlantic Ice-Age faunas, only taking longer to occur.

Casting back over the evidence, the very patterns of extinction, and some of the indications of prevailing physical conditions on the earth for the major extinction events, there is the beginning of a new consensus that global climate change—specifically, global cooling—is virtually invariably involved.

Cooling is strongly implicated in the Upper Ordovician double wave of global mass extinction. The late Devonian extinction now has likewise been linked to global refrigeration. Even the Late Permian, with its greatest extinction of all time, and traditionally associated with global warming, now seems to have been associated with global cooling.

Widespread salt deposits, formed when evaporation exceeds input of fresh water into a marine setting, are usually associated with warm conditions (high temperatures speed evaporation). But as Steve Stanley points out, salt deposits indicate dry conditions, and not necessarily warm conditions. Because direct evidence for continental glaciation is lacking in the Upper Permian, geologists have long assumed that the glaciers of the Upper Paleozoic actually were melting; as we have seen, recent evidence from Siberia seems to refute this notion, and the movement of continental plates off the South Pole by no means indicates that major polar ice caps were absent.

Yet there is a remaining problem: There is indeed evidence of major continental glaciation earlier in the Upper Paleozoic, with no concomitant mass extinctions. Stanley counters this with the claim that not all glacial expansion will trigger mass extinction, if the effects of the initial refrigeration already produced extinction shock waves through the biota, as had already happened in the Upper Devonian. Yet why, then, would presumed continuation of global cooling in the uppermost Permian all of a sudden produce the greatest extinction of them all? It is clear that not all periods of glacial expansion produce equally large-scale mass extinctions. We can presume that critical thresholds have not been reached in such instances, but we must also admit that the relation between extinction and global climatic change, usually cooling, while supported by a growing mountain of evidence, still needs a great deal more study.

Of the remaining mass extinctions that we have so far considered, only the Triassic lacks convincing evidence of an underlying association with global cooling. Even the bolide scenario (and, recall, that something of that nature must have happened to have produced the iridium spike) at the K–T boundary evokes global cooling (along with wildfires and occlusion of sunlight, blocking photosynthesis) as a direct effect and presumably as an important cause of extinction. Indeed, the bolide scenario is an excellent example of a presumed multiplicity of causes contributing to the waves of extinctions below, at, and above the boundary.

But there is that second, competing, earthbound cause of global mass extinctions yet to be considered. That is simply loss of habitable area. Most of the data supporting this notion come from the sea, specifically when sea level drops so low as to eliminate the shallow-water seaways of continental interiors, as well as along the marginal shelves surrounding the continents. There is one major strike against this notion: In marked contrast to global climatic change, regression of the seas, whatever their effect on marine creatures, will have the opposite effect on terrestrial organisms.

Loss of seaway means creation of terrestrial habitat. The ramifications are many: For example, when Central America last became dry land some 4 million years ago, all of a sudden there were faunal connections possible between North America and South America. Animals started to migrate up and, especially, down that newly created corridor right away. At the same time, there was no longer a direct marine connection between the western Atlantic and the eastern Pacific, and some new marine species have already appeared, as a direct reflection of the sundering effects of that isolation.

The idea that loss of habitable area is directly involved in extinction is a very old one. Paleontologists have long known that disappearance of species is often correlated with temporary retraction, or at least restriction, of marine habitats over the continental interior. Raymond Moore, the dominant figure of midcentury American invertebrate paleontology (and the man who started up the monumental compendium, the *Treatise on Invertebrate Paleontology,* which has served as the initial data base for much of the recent research on mass extinctions) pointed out that with the shrinking of the inland seas the invertebrates would all be crowded together on the continental shelves. This scenario directly invokes a picture of heightened competition simply for living space, and for the energy resources needed for each and every invertebrate animal to exist. Such times were bound to provoke extinction of species. And the longer and the more widespread the period of regression, the more devastating the potential extinction events could be. There could well be a sliding scale from minor to major effects of habitat.

I have seen a compelling example of precisely this effect in my own research. It is on a small scale, involving the disappearance and presumed extinction of a single species. The scenario does entail relations—ecological as well as evolutionary—with a descendant species. The small scale of this example suggests background extinction, but might also suggest that effects can be magnified into larger-scale extinctions, perhaps blurring somewhat Dave Jablonski's useful distinction between background and true mass extinction.

The trilobite *Phacops milleri,* derived from African stock, took up residence in eastern and central North America at the dawn of the Middle Devonian, perhaps 380 million years ago. Soon thereafter, it apparently gave rise to the descendant species *Phacops rana.* The two lived side-by-side for 2 to 3 million years: *P. milleri* lived on the limy bottoms of what is now the American Midwest, while *P. rana* claimed the sandy and muddy bottoms of what are now the Appalachians and the Allegheny plateau. You never find them together during this interval.

Finally, though, the seas of the Midwest dried up, and of course the entire marine biota disappeared. In the eastern seas, though, the sea stayed put, and the organisms apparently suffered no ill effects. In due course, the seas returned to the Midwest and virtually all the species that had been there before returned as their habitats were restored. Presumably, the organisms of the older midwestern sea all perished, and the new recruits were derived from their kin whose stocks persisted in the East. But *P. milleri* never made it back: Missing and presumed extinct, the inference is that it had nowhere to go. The seas in the East were already occupied by an ecologically similar sister species. So, when the seas came back to the Midwest, it was the eastern *P. rana* that came along, able to live there now that its sister species had succumbed to extinction.

I suspect this simple example is very general. More detailed study will probably show that many of the invertebrates—not just these two trilobite species—were actually different, closely related species, with one of the pair living in the shallow seas of the interior, the other adapted to the somewhat different ecological conditions of the eastern seaways. In any case, here at least is one concrete example where loss of habitat through regression seems very definitely to have been the cause of the demise of a species.

Ray Moore's young colleague, Norman Newell (already mentioned as the paleontologist who, nearly single-handedly, kept the study of mass extinctions alive from World War II until the explosion of work in the 1980s), was especially attracted to the loss of area hypothesis. Newell noted that the chalk deposits of the Upper Cretaceous, as well as the salt deposits late in the Permian, betokened periods of retreat of the seas from the interior regions of continents. Having receded, the seas quietly remained at their lowered levels.

Significantly, for both salt and chalk to accumulate, there can be no major influx of sands and muds from the continental interiors, and that seemed to indicate to Newell that erosion had lowered the levels of the continental interior drastically. And that implied loss of life-supporting nutrients for the marine creatures and (as imaginative paleontologist Robert Bakker has speculated) a decline of heterogeneity of terrestrial environments. Bakker ingeniously suggests that dinosaur extinction reflected more a drop in the rate of production of new dinosaur species (itself a reflection of increasing homogeneity of terrestrial habitats), rather than an increase in extinction rate

165

of dinosaur species per se—a clever way of relating sea-level change to extinction effects on both land and sea.

We have already seen how lowering of sea level is the very evidence cited to support global cooling and that many prefer cooling to habitat loss as the actual cause of extinction. The habitat area loss postulate received a strong theoretical shot in the arm in the late 1970s when paleontologist Tom Schopf (founder of the journal *Paleobiology* and the very personification of a new breed of theoretically inclined paleontologists) put a relatively new branch of biological investigation together with the data on mass extinction of the Upper Permian.

The theory of island biogeography, briefly encountered at the end of chapter 2, was proposed in a 1967 monograph by famed ecologist Robert MacArthur and ant systematist and ecologist E. O. Wilson, who soon thereafter wrote his *Sociobiology: A Synthesis,* causing quite a stir and essentially founding yet another branch of active biological investigation. Wilson is now one of the world's leading spokesmen for conservation and the importance of systematics research in the fight to understand and help preserve the world's threatened biotas.

MacArthur and Wilson established the close correlation between the size of a habitable area and the number of species found living on it. Small islands have fewer species than larger islands. And the relationship (with the inevitable, yet largely explicable exceptions that prove the rule) is mathematically linear. The conclusion seemed irresistible that if larger habitat areas have more species, then if areas were diminished there should inevitably follow a crash in numbers of species, echoing in mathematical terms the relation informally noted not only by paleontologists but by students of the living biota as well, for many, many years.

Yet the relationship between areal habitat extent and extinction has proven difficult to document in detail. Steve Stanley is a particularly persistent critic of the idea, noting that many analyses (such as Schopf's application of island biogeography theory to the end of the Permian crisis) utilized inadequate data and that not all regressions lead to mass extinctions. But I think it should be plain to all that the two effects can, and probably must, work together. And this is the form of an integrated theory of mass extinction that makes the most sense to me:

In any scenario relating climate change, especially global cooling, to mass extinction, we are dealing as well automatically with visions of habitat shrinkage and eventually habitat loss. Extinction, remember, by definition will not occur as long as organisms within a species can find (can recognize) suitable living conditions. They must do so in sufficiently large numbers to fit their required densities for reproduction to continue. Remember ecologists Dan Janzen and George Steven's living dead: Species, especially in the tropics, appear to be more far-flung than their actual, functional distributions really are. Lone outposts, recognizing living space, but in densities so low that reproduction will not take place, are functionally, genetically dead. Extinction occurs when a species is reduced in numbers below a limit that renders all remaining organisms members of the living dead.

The tropics are invariably hit harder than the temperate latitudes in mass extinctions because areal extent of true species ranges is far more restricted, on average, for each species than in higher latitudes. Because of narrow adaptation, there is simply less habitat area that will prove to be suitable to existing species as climate change hits. This is as true on land as in the sea. It will be far less likely for a tropical species to move to recognizable habitat than it would be for the species of the higher latitudes. Tropical species face both habitat area reduction and downright loss, and are far less likely to find escape routes. The entire tropical realm shrinks—direct habitat area loss—in the face of global cooling. The cooling hypothesis relies heavily on a particular form of habitat area loss to work.

Most scenarios of habitat area reduction invoke direct competition among species crowded together as the mechanism actually leading to severe extinction. This need not be the case—or, rather, the competition can be more implied than real. It is true that a species that might be able to gain a foothold in an area might be prevented from doing so if its prospective niche (the kinds of things their organisms would be doing to make a living) was already pretty much occupied by another species (whether closely related or not). That is competition by implication: The habitat is recognizable, but somebody else is already occupying it.

Just as likely, though, is the scenario suggested by the living dead. There will be other marginally acceptable habitats, but none good enough to support population densities required for true, full occupation: economic survival and ongoing reproduction. I have always thought that the extinction of the trilobite *Phacops milleri* reflected bad luck. The inland seas dried up, and the potentially suitable escape habitat was already occupied by *Phacops rana*. That *Phacops rana* was able to spread over the continental interior once the habitats were restored is just the reverse sort of situation: Suitable habitat not already occupied by a species with a closely similar niche became available.

Environmental change begets habitat change. Habitat change at any one place simply means shrinkage, perhaps to the point of total loss, of one set of living conditions replaced by another set. Extreme examples include ice fields covering tundra, or seas flooding continental interiors. Other, more subtle examples are, of course, much more common. Global climate change—cooling, but potentially also warming—shifts habitats around, shrinking some, expanding others. Survival of species is a matter of continued habitat recognition: There is a critical limit to habitat recognition depending on the abilities of organisms within a species to move, and on the width of their niche adaptations. Habitat area is a critical variable, but in turn is governed by general climatic factors.

Steve Stanley is probably right that marine regression alone (if, say, it were caused simply by seafloor spreading and not by glaciers indicative of global cooling) most likely is not the pure underlying cause of the mass extinctions of the geological past.

But changes in habitat distribution, which themselves involve changes in habitat area, as it interacts with the habitat requirements of organisms, is what determines survival and extinction. Global environmental change causes mass extinctions, but works through its effects on the relation between niche and availability of habitat.

Now we can turn to relatively recent times, see how these ideas fit the patterns of Ice Age extinctions, and evaluate the impact of that important added ingredient to the scene: *Homo sapiens*.

CHAPTER SEVEN

The Ice Man Cometh: Climate Change, Human Action, and the Great Pleistocene Extinctions

NATURE HAS NOT required the presence of our species, *Homo sapiens,* for its essential dramas to unfold. Life emerged at least 3.5 billion years ago. It evolved totally without reference to that style of existence we term *human*. And, as we have seen, life has come quite close to ceasing existence from time to time, as the eons have rolled by. Even looking to earthly causes—habitat loss, especially through climate change, being the main culprit implicated—there was no role for humans to play. There were no humans, and the biotic film clinging to the earth's outer rind has undergone its long periods of status quo, its bursts of exuberant diversification, and its traumatic throes of upheaval and extinction, wholly without our helping hand, or nearly so.

Though we are latecomers to the scene, there is plenty of evidence that humans—members of our own and other hominid species that have lived during the past 3 million years or so—have indeed had an impact on our fellow organisms. We are both a product of the natural forces of evolution and extinction, and an active agent that has already driven some other species to extinction and drastically altered the evolutionary histories of others. All of our forefather and collateral-kin species are already extinct. Some anthropologists even think that *Homo sapiens,* invading Europe 36,000 years ago, were the cause of the demise of the Neanderthals, sometimes interpreted as a subspecies of our own species, but in all likelihood a fully separate, yet closely related, species of hominid.

We have evolved. Our closest relatives are already extinct, and our species will, inevitably, also one day be extinct. We are not above nature; we are of it. But we are distinctly and peculiarly of nature, for we are, paradoxically, adaptively so specialized that we have emerged as supreme ecological generalists. We are eurytopes, that is, "suited to a variety of habitats." Though our eurytopy, our generalist nature, springs in large part from the fact that our species has come from a long line of physically and behaviorally generalized ancestral species, what really makes us generalists, able to penetrate and conquer so many habitats, is our specialized adaptation: our culture. We are conscious, sentient animals. With language, we communicate, and our entire economic and behavioral reproductive systems (diverse as they still are around the world) are complexly entwined in cultural conventions. We see nature—the physical world, and other species of animals and plants—in a way apparently no other organisms have ever done.

Our perception of nature is filtered through our cultures—from obvious devices like binoculars that help us see, to the more subtle conceptual tools we use to explore what nature is. To me, for example, species are real things, reproductive communities bound together by a shared fertilization system. To others, species are just collections of similar organisms, not especially real but rather artificial groups created by the human mind. We see nature, we relate to nature, through an age-old filter of human cultural history. We keep tinkering with that filter, and biologists are basically unified in striving for an ever-more accurate description of nature. But the fact remains that the very act of contemplating something sets the observer/

analyst perforce apart from the object under scrutiny; even biologists think that nature is something apart, something to be studied. We have, at least in modern, complex societies, become divorced from nature.

It wasn't always this way. Anthropologist Colin Turnbull has written eloquently on the closeness to nature that some African peoples still seem to have retained. I have always been somewhat cynical of the received picture of pristine peoples living in close harmony with nature, but Turnbull's picture of the Mbuti traversing the floor of the Ituri forest in the African country of Zaire, singing out "mother-forest" and "father-forest" as they go, is too compelling to ignore. The Mbuti call themselves "children of the forest," and believe that the forest gives them all they need: food, shelter, and clothing.

Culture is, in the broadest sense, our adaptation, our way of making a living in the context of local ecosystems. It may now be our cloak, that which keeps the cold cruel world from intruding too deeply into our lives, and simultaneously divorces so many of us (the urban dweller, sure, but also the rural farmer) from all but the most intrusive of nature's manifestations like the weather.

Our estrangement from nature is an understandable outgrowth, a by-product of our utilization of our very adaptation that allows us continued existence in the world. We have invented tools to allow us to hunt and later to domesticate and to farm; our species sprang from omnivores. We have even done something about the weather: We have clothing and elaborate shelter with heat, air conditioning, and indoor plumbing—all to keep the elements at bay. Our language-based cultural systems provide the tremendous advantage that information transmission need no longer be restricted to that which is passed along in the genes from generation to generation: We can tell each other how to do it, and each trick invented that helps wring an existence from the natural world is eagerly picked up, improved, and passed along.

It is not difficult to see how a process so beautifully geared to wresting a living from the natural world could become, fairly rapidly, ever-more efficient. Nor is it surprising that humans began to think of the process as a taming, even a conquest, of the natural world. If people have stopped short of declaring themselves gods, they have not been shy in thinking of themselves as cast in the image of God. Such, of course, can be an ennobling image, but it also represents a flat denial that we are of nature. The Judeo-Christian tradition exhorts us to seek dominion over nature. We have forgotten who we are.

There are those in Western technological society who insist that we really are not of nature, at least not so much that we need save "every species of squirrel" at the cost of economic progress (as George Bush's Interior Secretary Manuel Luhan said in 1990). There are others who feel that we are such a part of nature that the loss of every species threatens our own survival. These are tough issues. It has indeed been our cultural heritage for a very long time now that we virtually see ourselves as existing alone, divided into sometimes peaceful, sometimes warring, nation-states, and accompanied only by our domesticated pets and foodstuffs and the few species who seem to enjoy our company. We divide up the earth and buy land as if nothing else lived there, or at least as if their presence were irrelevant to ownership. Is this an accurate scenario? If so, can we survive as the dominant one of a handful of species? And, supposing we could, would we want to? Are we still more of nature than we think that such a monomorphic, unvarying existence would enable us to thrive?

We cannot ignore these issues, but we have to approach them with some information. We need to see how the presence of humans fits in with patterns of diversification and, especially, extinction over the past 3 million years. We are of nature, and our cultures have mushroomed to the point where we now tend to see ourselves as above nature, independent of it and no longer reliant on it or subject to its whims. How did we get that way? Has our desire simply to live, to make it in the natural world, ever led to extinction in the past? Or is it vainglorious of us to think we have ever really conquered nature at all? We need to come up with some strategies to minimize at least our own added contribution to what is, after all, a fundamental aspect of life on earth: extinction.

In the Beginning: Early Hominids of the Plio–Pleistocene

Large animals in big numbers invariably leave their mark on the local environment. The bigger the animals, and the relatively greater their numbers, the stronger their environmental impact: Elephants nearly wrecked Kenya's Tsavo Park a decade ago, when drought conditions reduced the edible vegetation below levels sufficient to sustain the existing elephant standing crop. Mountain gorillas, now so familiar through film and especially the books of Diane Fossey, likewise leave a trail of vegetal destructive chaos behind them: Gorilla troops are small, but so are their territories. Yet there is, on the whole, balance in such situations: Population size is determined by available and reliable (meaning renewable) resources. Over the eons, the interplay between reproductive biology and the exigencies of making a living determine characteristic patterns of distribution, and economic and reproductive behavior that fit organisms in a species to the nature of their habitats.

There is no doubt that, to a first approximation, hominid history over the past 3 million years is a story of expanding numbers: Body size has increased, and so has the population size of what started out to be rather few and widely scattered bands. The overall impact that humans have had on the physical and biological environment at base reflects the simple facts that we ourselves are large mammals and exist in devastatingly large numbers more typical (in the mammalian world) of rats and mice (and for the most part, only those species of rats and mice that have become tramps, roaming the world as hitchhikers and opportunists taking advantage of human existence).

The hominid story really does look like a march toward literal realization of the biblical injunction to "have dominion" over all other species.

But the hominid story is much more than a geometrically mushrooming cloud of population expansion. It is a story, as well, of speciation and extinction, and of exploitation of a variety of different ecological niches, as various of our progenitors and collateral kin have found themselves as parts of a large array of different ecosystems.

Charles Darwin, noting that most great apes live in Africa (and assuming that chimps and/or gorillas appear to be our closest relatives), naturally concluded that Africa must have been the cradle of humankind. Even after Darwin had convinced the thinking world that life must have evolved by natural processes, people had a hard time accepting the existence of bona fide human fossils. A German anatomist, Friedrich Mayer, for example, a dominant figure in late-nineteenth-century science, pronounced the skeletal remains found in a cave on Gibraltar in the 1840s to be the bones of a "rickety Cossack" who had presumably crept away from his confreres to die alone during battle. It is actually a fine specimen of a Neanderthal.

Somewhat later, the Dutch physician Eugene Dubois, frustrated in his desire to take Darwin up on his prediction and prospect for fossil hominids in Africa, settled instead for a stint in the Dutch Army in the East Indies. It was Dubois who found the skull cap and femur (upper leg bone) of what he dubbed *Pithecanthropus erectus*. The rugged skull cap roofed a brain much smaller than ours, hence *Pithecanthropus*, meaning "ape-man." But the leg bone was so much like our own that it undoubtedly belonged to a fully erect, bipedal creature. Skeptics insisted that Dubois had merely found the bones of an extinct ape, and Dubois himself pronounced his fossils (which he had meanwhile hid under the floorboards in his dining room) as the bones of a giant Ice-Age gibbon. We now know these bones as the first of many known of the wide-ranging, long-lived species *Homo erectus*—our Mid-Pleistocene progenitor.

I am not suggesting that all early discoveries of genuine fossil hominids were universally dismissed. Most biologists realized that the very idea of evolution must embrace humans as well, or be robbed of any real meaning at all. Though metaphorically humans had long tinkered with the possibility of their own godlike affinities, it was nonetheless always known and never really forgotten that we are animals, and therefore undoubtedly we had evolved along with all the other animals, plants, fungi, protists, bacteria, and viruses. There simply must be human fossil remains, though there has always been this certain reluctance to acknowledge them for what they are when first they come to light, a pattern that persisted until well into this century.

The last celebrated case of ancestor rejection came in 1924, when Raymond Dart, a young English anatomist teaching at the University of the Witwatersrand in Johannesburg, opened a box sent to him by a quarry foreman. Fossil baboons and other mammals had been known from the limestone works at a place called Taung, and Dart had an arrangement with the foreman to send along crates of fossils whenever supplies accumulated sufficiently. Dart pried open this latest box when he stole a few moments from the preparations for a wedding party. When he removed the lid, one specimen leapt immediately to his anatomist's eye. There was a little skull, with the outer bones broken away, revealing the cast of the brain. And Dart saw at once that the brain was human—and not the brain of a monkey. Careful work in the following days revealed the face of a child.

Human brains differ from ape brains in subtle but consistent ways. Dart had recognized features known only from humans among the living primate species. But, once again, a claim for existence of a primitive member of our own phylogenetic pedigree was rejected, this time by an establishment that, though fully geared to accepting the existence of hominid fossils, nonetheless felt Dart's material far too primitive, far too apelike, to truly be one of us, however far removed.

There were other factors underlying the rejection, apparently including professional rivalry and the belief that the "missing link" had already been found in the unfortunate *Eoanthropus dawsoni*, the fraudulent Piltdown man that had taken in a number of the leading physical anthropologists of the time, including the very men who had trained Dart and were now rejecting his claims of humanity for what he had named *Australopithecus africanus*. We now take this species to lie very near to the base of our evolutionary tree. And though, at least by Dart's time, there were all sorts of reasons to reject someone else's fossil candidate for human ancestry (it still goes on, as anyone familiar with the Richard Leakey–Donald Johanson jousts well knows!), the age-old and deeply ingrained precept that humans somehow are set apart from nature has left a bold signature on the checkered history of our collective attempt to understand human evolutionary history. Even professional anthropological paleontology has been slow to relinquish culturally ingrained visions of our place in nature.

It is notorious that there is no single interpretation of the hominid fossil record. Particularly absurd are the oft-repeated calls for more data: As everyone knows, every time a reasonably informative specimen is unearthed (more often than not, in East Africa these days), a cacophony of voices pushing rival interpretations assaults the ears. Particularly silly is the claim that discovery of a single specimen forces everyone back to the drawing boards, to throw out all other interpretations and completely revise the entire scheme of hominid phylogeny. But the cry for more data, when occasionally answered with fresh discovery, only brings a fresh spasm of disagreement.

We frequently do not appreciate the fact that with all the intensive post-World War II effort, spearheaded by the senior Leakeys (who had started work long before the war, but who had found no hominid fossils until the 1950s), the human fossil record is rapidly becoming respectably dense. The ratio of anthropologists to number of bones available for study is still far greater than it is for most other mammalian groups. We might have our problems understanding our history, but we persist, simply because the work is so fraught with implications for understanding who we are, how we fit into the natural world in the past, and what this all means for seeing ourselves vis à vis the rest of the natural world right now.

There are some fairly definitive things that can be said about the age, nature, and patterns of evolution and demise of extinct hominid species. (I use *hominid* here in a restricted sense, referring only to *Homo sapiens* and members of our lineage that appeared since the split between our ancestors and those of modern great apes some 4 or 5 million years ago. Properly speaking, by most anthropologists' reckonings, we, our fossil progenitors, and the great apes all belong to the single Family Hominidae.) For one thing, one would expect that the further back in time you search, the more apelike human fossils are bound to be.

When biochemists Mary-Claire King and Allan Wilson published the startling statistic that modern *Homo sapiens* and chimps (the species *Pan troglodytes*) appear to share a whopping 99 percent of genetic information, they were merely confirming what comparative anatomists had known for centuries: We are very closely related to chimpanzees. The statistic continues to have great shock value in confrontations between evolutionary biologists and creationists—those who would deny our natural pedigree and affinity with nature, in preference for the biblical myths of supernatural creation.

Though creationists these days are conceding microevolutionary connections between all manner of species, it is humans they have uppermost in mind when insisting on supernatural creation. The 99 percent figure also reminds us that what we take to be such wide differences in appearance and intelligence, especially language use and all that implies for the development and maintenance of cultural systems, can spring from such a proportionally small base of genetic difference. In any case, all signs point to close affinities between ourselves and great apes, and the fossil record should reflect that very fact: We would predict that the earliest fossils of the distinctly human family tree should look an awful lot like apes, minus some of the special features that ape evolution has stamped on their side of the family tree.

And that is precisely what we find. Don Johanson's famous Lucy and her kin, known from 3- to 4-million-year-old sediments in Ethiopia (and also claimed to be represented by equally old fossils in East Africa), have a distinctly apish leer. Brain size in this species, *Australopithecus afarensis* (meaning southern ape from the Afar region of Ethiopia), is just about chimp dimensions—some 450 cubic centimeters, on average. Anthropologists Jack Stern and Randy Sussman have analyzed the foot bones of these fossils, concluding that they were still preeminently suited for grasping, suggesting very strongly that Lucy and her friends spent a great deal of time in the trees in the woodlands of her home territory. (There is legitimate debate here: 3.5-million-year-old fossils from Laetoli in Tanzania are also considered to belong to *A. afarensis,* and there are several famous sets of footprints left by fully bipedal hominids frozen in lithified volcanic ash at the site.)

Certainly, in terms of overall body size, these primitive hominids were no bigger than good-sized chimps. Technically in the human lineage, these creatures were functionally far more like savannah-dwelling apes (who could still resort to trees!) than anything we would recognize as distinctly human. Anthropologist Tim White, who along with Johanson analyzed and named the Ethiopian and Laetoli specimens *Australopithecus afarensis,* believes this species to have been the only hominid species living between 3 and 4 million years ago. Johansen and White think that *A. afarensis* was in fact the ancestor of all of the rather diverse array of hominid species that was to follow, a reasonable rendering of the picture as it has taken shape so far.

The picture becomes a bit richer and more complex when we look at the next time slot, beginning about 3 million years ago. Here we find Dart's Taung baby, plus the other South African specimens discovered later and put, along with the Taung child, into the species *Australopithecus africanus*. All appeared to live sometime between 3 and 2.5 million years ago. (The Taung baby, never precisely dated before its deposit was essentially quarried away, seems to have been a bit younger, perhaps around 2.25 million years old.) These creatures weighed about 100 pounds and had brains that averaged around 450 cubic centimeters.

Now, the salient feature about these little hominids is their utter lack of any truly remarkable, distinctive characteristics! They have some anatomical features that mark them as a bit advanced over *A. afarensis*, but otherwise they seem perfectly generalized little hominids—primitive versions, in fact, of ourselves. Their slightness of build, in sharp contrast particularly with their later cousins, marks them as gracile. We are another example of a gracile hominid: light skeletal construction. This body type implies ecological generality, at least in contrast with some of the robust and ecologically very specialized later members of the hominid tree.

Once interpreted as rather nasty little hunters (an image portrayed by playwright Robert Ardrey in his *African Genesis*), we now see these gracile *A. africanus* individuals more as the hunted than as the hunters (though they no doubt ate small animals when they could catch them). The South African sites are accumulations of bones in caves that had appeared in ancient limestone. The bones are welded into a tough breccia that consists of broken fragments of rock and bone cemented together by redeposited lime. Paleontologist Bob Brain, who has devoted so many years to the excavation and analysis of South African cave fossils (particularly those at Swartkrans), has found that the bones of these early hominids accumulated, along with those of many other species, as they were dropped in from above—presumably by leopards lying up and consuming their kill in trees, which even today tends to occur around the openings to these subterranean caverns.

Rather than hunting, *A. africanus* individuals seemed to have lived mainly by gathering fruits and other plants. Although Dart thought he had evidence of a bone-tooth-horn (ostedontokeratic) cultural kit bag of tools (supplying the evidence Ardrey used to build his killer ape thesis), Brain thinks that all these tools are likewise just the remains of leopard meals. Stone tools were not to be developed until just under 2 million years ago.

Then, sometime around 2.5 million years ago, things changed rather radically all over the world, and certainly in Africa. The best African data are from the sites in the stratified beds of East Africa—the famous Olduvai Gorge in Tanzania and Richard Leakey's beds near Lake Turkana in northern Kenya, near the Ethiopian border being particularly noteworthy. These beds are well dated: Lava flows and volcanic ashes can be dated directly radiometrically. What's more, the earth's magnetic field (for reasons not fully understood) periodically loses strength, even dropping to zero. When the field builds back up, it sometimes adopts a reversed polarity, the North Pole becoming, say, negative and the South Pole becoming positive, whereas just the opposite had prevailed before the field lost steam. These old pole polarities are faithfully recorded in rocks with high iron content: There is a trace of the earth's regnant magnetic field when the rocks were forming still present millions of years later. It is possible to match up times of positive and negative polarity, thus really getting a good idea of what was happening on land and sea at the same time, which is vital for an accurate assessment of environmental and evolutionary events.

With such well-correlated sites, it becomes clear that about 2.5 million years ago there was a pronounced episode of global cooling. We are still well within the Pliocene epoch at this point, which is a bit confusing because we tend to associate the

Ice Age with the later Pleistocene Epoch (which did not begin until about 1.6 million years ago). But climatically, the Ice Age really began in the Pliocene, 2.5 million years ago, bringing with it a radical reorganization of the environment. And that occasioned the extinction of many taxa and the subsequent proliferation of replacement species. Paleomammalogists tend to think of the event as the beginning of the modernization of the world's mammal fauna. It was an event of no mean implications, as well, for species within our lineage.

Paleontologist Elisabeth Vrba had this 2.5-million-year-old environmental event firmly in mind when she framed her turnover pulse hypothesis, briefly encountered in the last chapter. The African (especially in the east and south) response to the worldwide lowering of temperatures was a loss of tropical forestation and the development of extensive woodlands occurring as islands within a greatly expanded open grassland and savannah habitat. The result was extinction for many species of antelope and other mammals. They were followed by species that either migrated in or were newly evolved: In either case, these species were able to take advantage of the newly available habitats. And along with all the other mammals, as well as other vertebrates, invertebrates, and plants, caught up in this bout of climate change, the early African hominids were also severely affected. The event changed the course of human history.

Australopithecus africanus disappeared about 2.5 million years ago and does not show up on the younger side of the event boundary, though the Taung baby may in fact be a bit younger than 2.5 million years—the dating is uncertain. This gracile, ecologically generalized species, with small body size and undoubtedly living in small-sized and probably fairly far-flung bands, had no hand in causing these events. Rather, just as they were victims of leopard predation, and not the active hunters Ardrey thought them to be, they were also the victims of ecological collapse: disappearance of familiar habitat, and loss, no doubt, of familiar food sources on which they depended.

But we are here, so loss of the single known hominid species between 3 and 2.5 million years ago clearly did not spell the end for hominids. Right after 2.5 million years, something rather new appears on the hominid scene. With massive jaws and skull ridges to support the huge muscle masses needed to move those jaws, the robust autralopithecines appear. In a way, they seem more primitive than the earlier *A. africanus:* The massive build of the face and the huge teeth bespeak something more along the lines of a gorilla than a human. *A. africanus,* remember, was gracile, a creature more easy to see ourselves in than the bigger, more massively robust forms that show up later, first in East Africa and soon thereafter in South Africa.

These robust species were all clearly adapted to chewing up very gritty tubers along with nuts and other vegetative food-stuffs. *A. africanus* may well have done a bit of scavenging to supplement its vegetarian diet, but it is presumed that these huge bruisers, like mountain gorillas (whose diet is almost solely bamboo), consumed only tough plant materials. There are now at least three species of these specialized vegetarians known (remember that ecological specialization seems to beget higher rates of both speciation and extinction).

The earliest species (known from a fairly recently discovered specimen referred to informally as the "black skull") has been named *Australopithecus aethiopicus.* The next oldest is the East African species, first called (by Louis Leakey, whose team discovered it in 1959 at Oldovai Gorge) *Zinjanthropus boisei* and later modified to *Australopithecus boisei.* Youngest among the robust species (but the first to be discovered) was *Australopithecus robustus,* a South African species. *A. robustus* and *A. boisei* lived to about 1.5 million years ago. Because they represent a distinct lineage (and have nothing to do directly with our own evolution), they become our phylogenetic collateral kin, and deserve to be so designated with their own separate generic name, *Paranthropus,* first proposed years ago by the prescient anthropologist John Robinson.

If these specialized herbivorous species had nothing to do

with our own eventual emergence, what did? The famous species *Homo habilis* ("handy man"), named by Louis Leakey primarily on the basis of a fairly modern-looking hand discovered at Oldovai Gorge, is the candidate. Appearing just a little more than 2 million years ago, *Homo habilis* is considered to be one of our direct ancestors. True, it is hard to tell some specimens of *Homo habilis* from *Australopithecus africanus*. Later on in its evolution, it is hard to tell where *Homo habilis* stops and *Homo erectus* begins.

Eurytopic species tend to persist for relatively long periods of time, and may accumulate a fair bit of change along the way. But, as we have already seen, we are rather a strange kind of eurytope: It was the primitive, unspecialized gracile forms—the presumably ecologically generalized forms—that led to ourselves. Yet the hallmark, quintessentially human adaptation is culture: the very feature that has allowed us to make such inroads against the environment, to come so close to achieving true dominion over the other beasts of the fields.

Sure enough, it is when *Homo habilis* appears that we start to find the oldest of manufactured tools—very simple, even crude, choppers fashioned from cobbles. Anatomically, *Homo habilis* represents a tremendous step forward in that very crucial aspect: brain size. In one spectacular specimen discovered by Richard Leakey at his site (Koobi Fora) at East Turkana, the brain size is approximately 780 cubic centimeters. Ours is around 1,400 cubic centimeters, and some Neanderthals evidently had even larger brains, but 780 cubic centimeters is a giant leap beyond the 400- to 500-cubic-centimeters capacity of the earlier australopithecines.

With *Homo habilis,* the basic features of human existence that we tend to think of today began to take shape. The species was not long-lived (and little is known of what may have occasioned its demise), but *Homo habilis* did give rise to the highly successful *Homo erectus*. For the first time, humans broke out of Africa, to colonize much of Europe and Asia. We are, by now, firmly within the Pleistocene: *Homo erectus,* first known from (and undoubtedly originating in) Africa, appeared around 1.6 million years ago—evidently a child of the next cooling spasm, the one that officially rang in the true Ice Age (even though that event apparently had much less far-reaching consequences than the Pliocene cooling spasm at 2.5 million years ago). It was not until some time later that *H. erectus* migrated out of Africa; the youngest specimens known, from the famous cave (Zhoukoudian) near Beijing, lived more than a million years later than the oldest known African specimens.

For most of the Pleistocene then, from 1.6 to perhaps 0.3 million years ago, there was but one hominid species on earth. Confined entirely to the Old World, this species has left its stone tools (many times more sophisticated than the crude Oldowan choppers) all the way from Capetown to central Java and many places in between, from North Africa, through western and central Europe. Glaciers were coming and going, as were the wetter pluvial periods of Africa (these days once again considered to occur at the same time as glacial growth in the higher latitudes; the connection between increased rainfall in the tropics and high-latitude glaciation, if there is such a connection, is by no means completely understood).

Homo erectus was beginning to show the toughness and flexibility of a true eurytope. Anthropologists argue over the amount of evolutionary diversification this species showed as it spread over a huge territory and persisted for well over a million years of the Pleistocene. To some, there is more regional diversification in stone tool types than there is in physical conformation, but others argue that such critical features as brain size in fact do show progressive increase through time. Sample sizes are not optimal (we need more data!), so the average cranial capacity of the late Chinese specimens seems significantly higher than the million-year older African specimens—that is, to some anthropologists. Others, pointing to an early African cranium with one of the highest cranial capacity values obtained for a specimen of *Homo erectus* of any age, claim that the differences in brain size between the earliest and latest specimens are in fact insignificant.

What is quite definite is that *Homo erectus* was a highly successful species whose basic cultural and physical ecological adaptations managed to remain remarkably stable in the face of prodigious climatic oscillations. Temperatures dropped and climbed and other species became extinct, yet *Homo erectus* just kept on going. If not exactly oblivious to environmental change, there is every reason to suppose that, with this species, the ability to step away from such vicissitudes—to gain some mastery over a range of environments, to have the flexibility to cope with altering conditions, trademarks that we have within our own species—was definitely beginning to emerge. Those cultural and physical adaptations remained stable because they worked. And they worked, even when the environment swung through its wild oscillations, because their very essence is creative flexibility. People were beginning to do something about the weather, not to change it but to deal with it and not fall victim automatically to its effects.

Homo erectus undoubtedly was a meat-eater, though details of hunting ability are sketchy. It used to be supposed that later populations had mastered fire, presumably added indication of a meat-eating habit. The attribution of fire, it turns out, was based on charcoal horizons within the caves at Zhoukoudien, and fires are now thought to have been natural rather than humanmade.

With *Homo erectus,* our lineage began to get into environmental mastery. Yet clearly this species, far-flung though it was, still lived in truly small population clusters. There is no indication of mass killing of prey animals, nor is there any other reason to suppose population size was even approaching the limits of resources. Flexibility was the key, with different populations in different places, at different times, faced with vastly different combinations of edible resources. There is no indication that *Homo erectus* was anything more than an unusually

187

clever participant in any of the myriad Pleistocene ecosystems of which it was a part.

Humans were still parts of local ecosystems then. Our flexibility showed up more in terms of survival and evolutionary implacability. *Homo erectus* survived relatively unscathed and unchanged for well over a million years. Flexibility, too, was manifested in the range of environments in which humans could exist, even flourish. There is no evidence that *Homo erectus* had any direct effects as causal agents of environmental change or extinction in the Lower and Middle Pleistocene. We were still of nature, and if nature had a little less control over us than it had in the past, we still had made only a small dent in it.

Climate Change and Extinction in the Pleistocene

Signs of the presence of glaciers go back as far as the Upper Precambrian. And we have encountered them particularly in the Upper Ordovician, less certainly near the end of the Devonian, and especially in the Upper Paleozoic. There is some thought that global cooling affected late Mesozoic life, though its cause is unclear: It may reflect the direct effects of short-term bolide impact. There was a cooling pulse again in the Oligocene. And, at 2.5 million years ago, still in the Pliocene, the coldest pulse of the entire Cenozoic up to that point had tremendous impact on rearranging habitats, driving some species to their presumed extinctions (they never reappear as Lazarus taxa) and paving the way for new species. Some evolved, and some merely came in from similar habitats elsewhere. This big drop in global temperature, with all its consequent dramatic effects, really signals the beginning of the Ice Age, even though the Pleistocene, long synonymous with Ice Age, didn't officially begin (by traditional geological agreement on boundaries) until 1.6 million years ago.

Glaciers, of course, are an effect, and not the cause of global cooling. What causes the cooling itself is still a matter of some debate, though some general principles are clear enough. The earth has two sources of energy: endogenous and exogenous. In other words, the earth manufactures its own energy through radioactive decay deep below the surface. Recall that if you go down any cavern or mine adit (with or without a canary), the initial cool dankness quickly gives way to warmth—a warmth that only increases the deeper you go.

Energy, in the form of heat, is constantly streaming up through the earth's crust, only to be dissipated as it reaches the surface. Heat flow is generally so negligible that the direct effects on climate, locally and when totaled, even globally, are slight. Exceptions include areas of concentrated heat flow, such as the deep-sea vents and terrestrial thermal springs that construct dramatically unusual microhabitats: The heat at the deep-sea thermal vents is captured by a sulphur-based metabolic pathway in bacteria that form the very base of the food chain at depths far too great for photosynthesis to occur.

There can even be direct and truly global effects of concentrated heat flow: When Krakatoa detonated in 1883, soon afterward sunsets were colored in the northern hemisphere, lowering global temperatures and apparently accounting for the "year without summer" in 1888 that not only produced the famous Blizzard of '88 in New York but also snow storms in July, temperatures far below normal, and general crop failure.

But such volcanic events themselves are not supplying energy to the surface environment: They are, instead, blocking energy inflow from the earth's other major energy supply, the one most directly important to surface temperatures and the existence of life on this planet: the sun. Ash disseminated in the stratosphere reflects sunlight, bouncing much of it back into space, and severely affecting the amount of solar energy able to reach the surface. Temperatures drop, rather than climb, at such localized and concentrated volcanic outpourings of endogenous energy.

There is another, albeit very complex, category of secondary effects that endogenous energy flow wreaks on surface climate.

Recall that it is internal heat that drives plate motions. The earth's basic weather patterns reflect motion within both the atmosphere and oceans. Such motions are determined to a great degree by the relative amount of land exposed and by the position of that land relative to the sea. This is most easily seen with oceanic circulation: Bring two continental plates together and the ocean no longer flows between them.

A relatively trivial yet sharply dramatic example occurred only 6 million years ago, when the African and European plates got so close together that what is now the Straits of Gibraltar were completely closed off. The Mediterranean became a vast inland and very salty sea. And because water loss through evaporation exceeds the amount of fresh waters brought in by the relatively few rivers (including the mighty Nile, Danube, and Rhone), the Mediterranean, over a period of about 1 million years, dried up! Vast salt deposits that form a very hard, continuous layer below the bottom muds testify to the event, a tremendous and direct effect of plate motions on the environment!

Yet other effects of plate motion and continental positioning are, if more subtle, even more profound. The Gulf Stream today leaves the western hemisphere after passing Bermuda, and, cutting a northeastward path across the Northern Atlantic, enables palm trees to grow in Great Britain, which lies as far north as Newfoundland does in the western North Atlantic. Oceanic and atmospheric circulation distributes solar energy, which otherwise would be more concentrated in the tropics and less felt in the polar regions than it already is. Anything that disrupts major patterns of oceanic circulation (which, in turn, determine atmospheric circulation to a great extent) is bound to have a strong effect on climate—and plate tectonics, driven by endogenous energies, is the ultimate determinant of the physical configuration of the earth's surface.

Indeed, it is plate tectonics, leading to the present-day configuration of continents and ocean basins, that has led to the great waxing and wanings of continental ice sheets in the northern hemisphere these last 1.6 million years. One key to the puzzle of global cooling and glaciation is the presence, over at least one of the poles, of a continental landmass. In the Ordovician, it was North Africa; today, Antarctica sits astride the South Pole. Even in dry climates with little precipitation, ice accumulates whenever seasonal periods of melting do not keep up with snowfall. The Antarctic ice pack goes back at least 5 million years. Cold waters from Antarctica flow northward, crossing the equator at great depth and effectively cooling the oceans, thus the atmosphere, in the northern hemisphere.

There is an entirely different class of features that greatly affects both the total amount of solar energy striking the earth and its distribution over the face of the globe. The earth, as everyone knows, does not stand straight up and down: Rather, its axis of rotation (giving the true, as opposed to the magnetic, north and south poles) is inclined at an angle of 23.5°. Add to that the fact that the earth's orbit around the sun is not perfectly circular, but rather elliptical, and we see why there are seasons, most marked in the higher latitudes and undetectable right along the equator. Right now, the northern hemisphere is pointing away from the sun near the farthest point away from the sun as the earth follows its elliptical orbit: That gives us winter in the northern hemisphere, where the angle of the incoming sunlight becomes quite oblique and the density of solar radiation is reduced. The opposite, of course, is going on in the southern hemisphere at the same time: The higher the latitude in summer, the later at night one can go outside and read a newspaper.

But this was not always the case, nor will it remain this way for long. The earth is literally a spinning top: Its axis wobbles, oscillating between 22° and 25°. And, just like the axis in a spinning top, it precesses, which is to say the tip of the axis describes a circle rather than remaining fixed at one point. The rates of these changes in tilt and wobble position, and how they relate to positions of the earth in its orbit of rotation, are known. Yugoslavian physicist Milutin Milankovitch calculated that the shift between 22° and 25° of the axial tilt occurs every 41,000 years. The wobble and change in position of tilt vis à vis the place in the elliptical orbit cause a shift in winters and summers every 22,000 years: When the tilt toward the sun coincides with the shortest distance from the sun, summers are at their very hottest. Add to this one more variable: that the earth's orbit is not invariably elliptical but sometimes approaches circularity, and there are the ingredients for a vast but regular and complex (yet interpretable) variation in solar energy reaching the earth's surface.

Over the past 1.6 million years, there have been about fifteen major periods of cooling and about fifty minor ones. Recent work in deep-sea cores has shown remarkable agreement between Milankovitch cycles and the swings of earthly temperatures during the Pleistocene. Though there have been but four or five major ice pulses in the northern hemisphere, each glacial advance is marked by minor episodes of warming (melting back) and cooling. But why have these Milankovitch cycles (which presumably have been operating ever since there was an earth revolving around the sun) only had such stark effects at isolated times in earth history: during the last 1.6 million years, and in the other notable periods of global cooling and glaciation?

The Milankovitch cycles yield regular oscillations between extremes in warmth and coolness in the northern and southern hemispheres. The trick to understanding the Pleistocene glaciations is that, during those periods of refrigeration in the north as the glacial fields were advancing further south and winters became extremely cold, the corresponding warmth of the south polar summer did not melt away the Antarctic ice cap. Effectively isolated by a circum-Antarctic current, there was no mass melting of the south polar ice cap as glaciers were advancing in the north.

That Antarctic ice cap, the product of plate motions that has positioned Antarctica directly over the South Pole, has itself been the cause of global cooling—a cooling effect that mitigates the warmer extremes of northerly summers and continues to refrigerate the globe even when summers reach their warmest in the southern hemisphere. Ongoing changes in oceanic circulation continue to modify the direct, straightforward effects of the orbital oscillations. It is plate tectonics and Milankovitch cycles and possibly periodic fluctuations of carbon dioxide in the atmosphere that have caused the wild climatic fluctuations of the northern hemisphere over the last 1.6 (or, actually, 2.5) million years.

Continental glaciers are something to consider. I am still amazed that a scant 18,000 years ago there was an ice sheet that reached as far south as Iowa, Illinois, Ohio, Pennsylvania, New Jersey, and New York. In places, the sheet was 3 kilometers thick! I have earlier remarked how different terrestrial habitats are when ice covers what now are plains or dense coniferous forests. And it is certainly true that, in the face of changes of such stupendous proportions, the first response of organisms is to flee.

As we have seen, even trees migrate in the face of an advancing glacier, or simply during climatic conditions too harsh for them to withstand: They move, generally southward, into refugia providing recognizably suitable habitat, simply by seed dispersal. Yet, given sufficiently harsh conditions, even the relatively more eurytopic species of the higher latitudes will tend to meet their Waterloo and become extinct. The Ice Ages are as renowned for their extinctions as they are for the evolution of their own, specialized organisms adapted especially to the cold, extreme climatic conditions of the glacial maxima. In many cases, it was precisely those huge, formidable beasts that both appeared and disappeared as the Pleistocene waxed and waned. A large body conserves heat and is a good defense against the cold.

The mammoths, mastodons, woolly rhinos, and giant bison, so conspicuous in our museums and dioramas of the past, were manifestly creatures adapted to life in and around the glacial fields of the frozen north. Imagine the tundra today taken to the *n*th degree. Paleontologists Dale and Mary Lee Guthrie have painted a compelling picture of what they call the *Mammoth Steppe,* a vast, supertundra so dry (and so cold: −60° F for long periods) that trees could not grow. Yet there was an astounding abundance of different mammal species able to survive—even thrive in—such extreme conditions. Clearly, the tremendous climatic fluctuations of the Pleistocene were as much a spur to evolution as they were a causal agent of organismic demise.

To a first approximation, the reasons for such marked faunal turnovers are clear: The increasing cold brought about the by-now familiar process of retreat toward the tropics. Steve Stanley, as briefly mentioned in the previous chapter, has documented this process in marine bivalves and has even demonstrated that the way southward proved to be easier for West Coast than Atlantic species alongside the North American continent. An unimpeded migration route and availability of suitable habitat once there resulted in far greater survival of marine clam species in the Pleistocene on the West Coast than proved to be the case in the East.

Climatic change, especially the cold snaps that send organisms looking for suitable refugia closer to the equator, creates expansion of some habitats at the expense of shrinkage of others. It is literally such habitat loss, mainly affected by shrinkage or even total loss of suitable habitat, that leads to extinction. Today, on the north side of the Himalayas, there is a vestige of that vast mammoth steppe once stretching from inland Alaska, miraculously free of glacial ice, right across Asia into Europe. True continental glaciers were largely restricted to east and central North America, and to western Europe.

There can be little doubt that, at least to start, humans had little (actually, in all likelihood, nothing) to do with the waves of extinction that occurred in the Pleistocene. We have already seen that the extinctions and consequent speciations that reflect the first cold pulse in the Upper Pliocene 2.5 million years ago greatly affected the tropical East African fauna and quite likely had a lot to do with the disappearance of *Australopithecus africanus,* and the first appearances of both the specialized robust australopithecines and species of the genus *Homo.*

Further to the north where no hominids (at least any belonging to our own branch of the family tree) had yet set foot, there was also prodigious faunal turnover 2.5 million years ago: The so-called Villafranchian fauna, known from Europe and with particular clarity from a profusion of fossils from the Siwalik Hills of Pakistan, saw the advent of the first true (modern) horses, as well as mammoths, bovines, and other large and typical Pleistocene forms (even though the Villafranchian began in the Pliocene, which still had 0.9 million years to run). The first glacial advance of the true Pleistocene seems to have had a less pronounced effect on the faunas as a whole, although the event does seem to be marked with the rise of *Homo erectus* in Africa.

There was a big wave of extinction about 0.9 million years ago. That is when the large mammals familiar to us from reconstructions of Los Angeles' famous La Brea Tar Pits (not to mention the incredible cave paintings dating from about 16,000 years ago, predominantly from Spain and France) first appeared. It also seems to mark the time when our lineage first trekked out of Africa. The appearance of *Homo erectus* in Europe and Asia seems to date from about the time of the advance of the great Elster ice sheet. While the glaciers waxed and waned, there was indeed evolution and extinction all the while, and some climatic events appear to have been sufficiently severe to trigger what is by now the familiar story of disappearance and extinction, followed quickly by the appearance of species new to the area as the habitat changed.

Some of those first appearances mark simple migration as organisms continue the search for recognizable habitat. But some of the species are truly new, newly evolved presumably in response to the availability of new ecological opportunities. But it is what happened to that last, final phase of the Pleistocene fauna—the woolly rhinos, cave bears, giant bison, mastodons, woolly mammoths, saber tooth cats (erroneously often called tigers), dire wolves, ground sloths, giant condors—that begins to raise questions about the roles humans may have played.

As the Pleistocene wore on, *Homo erectus* survived many a wild climatic swing. Times were warm; then times would be cold again. Culturally, at least, *Homo erectus* seemed not to have changed a great deal (to judge, that is, from the tools left behind). Physically, as we have seen, there are some paleoanthropologists who feel that the late Pleistocene *H. erectus* known from Asia are quite different from the earlier *H. erectus* known only from Africa. And, by Middle or Upper Pleistocene times (starting, that is, about 300,000 years ago), a number of fossils scattered here and there in the Old World suggest, at least to paleoanthropologist Ian Tattersall, that a number of species had evolved from the *H. erectus* stock in the heterogeneous Eurasian landscape.

What happened to *Homo erectus*? Extinction does not claim its victims because their cosmic time has come, and they have become archaic or outmoded. Nor do we any longer see extinction as the vanquishing of the old, the ancestral, the primitive, by their more advanced descendants produced in the course of evolution. Mammals did not evolve and then outsmart and outcompete dinosaurs, driving them to extinction. If anything, it was the other way around: Not until dinosaurs finally became extinct could the mammals (who had been around every bit as long as the dinosaurs—from the Upper Triassic through the entire Mesozoic) take over. As we have seen, it was only after the dinosaurs were gone that the mammals could take over, radiating, after a brief lag, explosively into a wide variety of shapes and sizes, reflecting their newly found opportunities to play the major roles in local terrestrial ecosystems.

Yet it is tempting to suggest that the appearance, finally, after nearly 1 million years, of one or more new, bigger-brained species of hominid was precisely what did in *Homo erectus*. We are, after all, ecological generalists—an ecological status bestowed by our generalized physiology and anatomy, but also, ironically enough, by our specializations: our large brains and our cultures that are enabled by our cognitive capabilities.

I believe that all members of the genus *Homo* (and probably gracile species of *Australopithecus* as well) were basically eurytopic: ecological generalists. And generalists do not, as a rule, coexist with close relatives. Eurytopes tend not to share niches. We did see that our ancient lineage split into two separate lines just after the big extinction event of 2.5 million years ago, producing the generalist species *Homo habilis* (our own ancestor) and the specialized offshoot, the strictly vegetarian robust lineage of australopithecines. Species of robust australopithecines and of our own genus *Homo* could live in the same general area, presumably because they were not directly in competition with one another.

But mobile bands of a gracile, eurytopic species tend to collide and compete with one another, as we know still goes on within our own species. It seems unlikely that two species of eurytopic hominid could coexist for long. Indeed, that is the picture that seems to hold a bit later, when we confront the disappearance of Neanderthals in Europe. Here, for *Homo erectus*, we simply lack the more detailed information that the fossil and archeological record gives us for the Neanderthal story. Indeed, anthropologists have not yet officially named any new species coexisting in Europe along with the remnant populations of *Homo erectus* known from Asia. But I cannot restrain the speculative guess that it was competitive pressure exerted by a rival (and descendant—shades of Oedipus!) species of hominid that finally did in *Homo erectus*.

In any case, by 100,000 years ago it was the Neanderthals who peopled the European theater during the last glacial advance. Somewhat stockier and classically more robust than modern *Homo sapiens*, Neanderthals are still considered by many paleoanthropologists to constitute merely a variant version of modern humans—the subspecies *neanderthalensis* of the species *Homo sapiens*. Once the source of the Alley Oop imagery of the stooped, cloddish cave man, the trend of more recent times has been to see Neanderthals, with their brains a bit bigger than our own, as advanced as modern humans, and cold-adapted to match the conditions of Upper Pleistocene Europe. But the anatomical evidence points to an inability of Neanderthals to form the same range of sounds necessary for fully human speech. That, plus the anatomical distinctions between these robust hominids and the gracile and completely modern *Homo sapiens* that show up so abruptly in European deposits of about 34,000 years ago, adds up (to my mind and, more significantly, the minds of Ian Tattersall, Chris Stringer, Eric Trinkhaus, and a number of other experts) to a convincing case that *Homo neanderthalensis* and *Homo sapiens* are two different species.

The old story, in fact, was that once humans spread out of the African crucible (now known to have been a feat first accomplished by *Homo erectus*) we simply evolved in place, throughout our recently expanded geographic range. Anthropologist Carleton Coon went so far as to suggest that the species *Homo sapiens* in fact evolved no fewer than three separate times at three different places! In his scheme, *H. erectus* populations gave rise to modern humans separately in Africa, Asia, and Europe (via the Neanderthals); the forerunners of the three great modern races of humans, in other words, directly reflect presumed racial differentiation already present among *H. erectus* populations. The usual view of species, though, sees them as reproductive communities that have distinct origins through a budding-off process from an ancestral species. By definition, species must have single origins; if *Homo sapiens* is a valid single species (and it most certainly is), and if it is distinct as a species from *Homo erectus, Homo sapiens* must have arisen in one particular locale from *H. erectus,* or from one of the possibly several species that may have evolved from *H. erectus* in the Upper Pleistocene.

Recently, both fossil and molecular evidence has accumulated that makes it quite plain that *Homo sapiens* did in fact have a single-point origin. Where? Right back in the crucible—somewhere in Africa. *Homo erectus* got out of Africa earlier, but it is almost certainly true that none of the Mid-Pleistocene hominids of Europe and Asia were ultimately to survive by giving rise to modern humans.

Allan Wilson, the same biochemist who, with Mary-Claire King, gave us the arresting figure of 99 percent genetic similarity between humans and chimps, led the team that has announced that the ancestor of *Homo sapiens*, literally the mother of us all, was an African woman living sometime between 100,000 and 150,000 years ago. Wilson was looking at the DNA of mitochondria, which are organelles inside cells responsible for energy conversion. His team was reconstructing the original sequence of the components of the ancestral mitochondrial DNA chain, trying to distill the common ancestral conditions that gave rise to the diversity known from different human populations alive today. We inherit our mitochondrial DNA strictly from our mothers. Hence the common ancestor of us all, as traced through mitochondrial DNA, was a woman—appropriately, "Eve." The reconstructions best match sequences known from present-day African mitochondrial DNA. We are all, ancestrally, Africans, twice over.

And quite recently Wilson's figure of 100,000 to 150,000 years agrees rather nicely with the 90,000+ date for anatomically modern humans recently found in southern Africa. Some of the dates for *Homo sapiens* in the Middle East (part of what is now Israel), long thought to be in the vicinity of 45,000 years, also have been pushed back—quite a bit, to around 90,000 years. And, as we have seen, we got to Europe about 34,000 or 35,000 years ago. In we came and out went the Neanderthals in what looks to have been an abrupt replacement that may very well have involved outright extirpation. For a brief time, it looks as if our species actually may have coexisted with Neanderthals in the Middle East. But our expansion into Europe has all the earmarks of an abrupt invasion, and there is no reason to suppose we got into Europe only after something else killed off the Neanderthals.

Neanderthals disappeared forever in a background extinction event: No other species is recorded as becoming extinct at the same time as the Neanderthals. My sense is that we were to Neanderthals what the brown tree snake was to the Guam rail. But the dynamics were different: No act of predation, or perhaps even of outright warfare, the two hominid species were simply in competition for the same resources, with the predictable result. One would come out ahead and eventually capture the entire habitat.

From Passive Victim to Active Agent: The Emerging Human Role in Extinction

The Mbuti of the Ituri forest are one of an ever-decreasing, small number of human groups that still retain the primordial human status as an integrated part of local ecosystems. Our entire species has had this status for the greater part of our history thus far. Even when the spectacular paintings of western Europe were executed on cave walls some 16,000 years ago, our species was still clustered in small aggregations, living off the land as hunters and gatherers. As we have seen, much of hominid evolution, certainly including some of the earlier episodes of extinction of hominid species, is tied in directly with general ecological events that simultaneously affected many other species as well.

As in all other species, our own history mirrors the impact of environmental change. But, as we approach modern times along the sweep of human history, we see a transition of *Homo sapiens* to an active role as direct agent of environmental change—first as a suspect, and finally as an undoubted cause, of the extinction of other species. This transition reflects our changing status vis à vis ecological systems generally. And our role as active agents of environmental change is intimately tied in with the geometric increase in human populations that has taken place over the last few thousand years.

Today *Homo sapiens* are all over the face of the earth. Populations have risen: They keep going up exponentially. Paleontologist Norman Newell, that voice in the wilderness in mid-twentieth century, virtually alone in his insistence on the reality of—and his zeal to identify the causes of—mass extinctions, has recently turned in a bit of elegant scholarship: Writing with paleontologist/statistician Leslie Marcus, the two have demonstrated an unbelievably tight correlation between the rise of atmospheric carbon dioxide and human population growth between 1958 and 1983. The correlation: 0.9985 out of a possible 1.0000! No set of empirical data ever shows such a tight correlation. Pitfalls aside of specifying cause and effect from mere correlation, there can be no doubt that the rise is caused by human activity—the fruits not only of the industrial revolution but the destruction by fire of much of the world's vegetation. (An infrared satellite photo of the earth at night shows some of the big population centers: The heat and light radiating from major cities of course shows up. But the light spots are mostly in Asia, Africa, and South America, and are simply forests on fire.)

Yet that was not the point Newell and Marcus were trying to make. Cleverly, they had in mind the flip side of the relationship: You can census the human population by simply measuring the amount of carbon dioxide in the atmosphere! (This only works globally, of course, because of atmospheric mixing; thus this technique offers no hope for rectifying the most recent U.S. census undercount!) We stand now at some 5 billion people and are still counting. Wars, pogroms, epidemics (tragedies all, scourges that we must continue to work to eliminate from human life) place the merest of dents in the geometric mushrooming of global human population.

Anthropologist Marvin Harris, writing not long after the first of the so-called energy crises engendered in the United

States from Mideast political problems, has written an engrossing account of human history from an energy-utilization point of view. He sees population growth not so much as a smooth, linear (if exponential) curve but rather as one loaded with conspicuous kinks if looked at closely enough. Populations, Harris reasonably enough assumes, are generally at or near their so-called carrying capacity, meaning that following Charles Darwin, the prevailing supposition is that population sizes are regulated by available resources. Unless deflected by other factors (disease, for example), populations are generally near their maximal size, given the resources and given the ecological adaptations of the organisms involved. Two species, with similar energetics requirements, may nonetheless exist at different optimal population sizes, depending on the nature of the way their organisms naturally exploit the available resources.

Harris simply thinks that various innovations in human adaptations have greatly expanded resource bases and have led, as a direct consequence, to great expansions of human populations. Take the invention of farming, which occurred in several nuclear areas, for example, the fertile cresent of the Tigris and Euphrates, the Indus Valley, the Nile Valley, and the region surrounding Mexico City (lacustrine, rather than riverine, in this case). Immediately population size increases, simply because reliable energetics resources (cultivated grain and domestic animals) beget not only a sedentary existence (i.e., as opposed to the partially or even wholly nomadic existence of hunter-gatherers) but an increased and regularly reliable resource base that is greatly in excess of what hunter-gatherers can count on. What is more, there is a maximum effective size apparent for the hunter-gatherer mode of life where bands typically share whatever is collected.

Farming does, of course, lend itself to small-scale endeavor (increasingly not a realistic option in the United States). But it is also the very basis of urbanization, as someone else, on a mega-scale, can be out there producing the food that urban dwellers can buy with the monetary fruits of their own particular labors from a different line of work. Culture-based technological revolutions, as Harris has stressed, typically hinge on innovations in energy-resource utilization, be it food production or the literal extraction of fossil and nuclear fuels from the earth, which keep on firing the machines of the industrial revolution. Our cultural inventions have periodically and rather abruptly expanded our resource base, and population size always followed shortly thereafter.

Our population increase has depended utterly on changes in our approach to economic needs. There is little question that it was the origin of sedentary life, supported by cultivation and animal husbandry, that changed the course of human economic adaptations for all time (barring nuclear holocaust scenarios that wipe out all but a technologically naive few who would start out a completely new phase of human existence). But before the critical change to farming, there was man-the-hunter. And there is an impressive pile of evidence that humans had much to do with the demise of the Late Pleistocene megafauna—the giant bisons, the elephants, the rhinos, creatures that roamed such places as far removed from Africa as North America, in some cases, as recently as a scant 8,000 or 9,000 years ago.

Paul Martin, an anthropologist at the University of Arizona, has devoted years to studying the effect that the peopling of the Americas had on the latest Pleistocene fauna. Though there are some supporters of the idea that humans crossed the unglaciated land bridge connecting Asia and western North America as early as 35,000 years ago, there is very little hard, convincing evidence of early settlement dating much beyond 18,000 years ago. The Meadowcroft Shelter, recently studied intensively by a team of archeologists and geologists, records the presence of paleo-Indians in what is now Pennsylvania about 18,000 years ago. But strong evidence of the presence of humans in significant numbers in the New World does not occur until about 12,000 years ago.

Particularly dramatic is one of the earliest archeological sites known for early humans in North America: Near Clovis, New Mexico, the remains of a bison were found with a spear point firmly embedded in its ribs. There is good evidence, as well, that early humans in the New World knew the same tricks used by earlier hunters of the Upper Pleistocene in the Old World: driving mammalian herds into blind corners and even over cliffsides, the better to dispatch them, if not at leisure, at least in relative safety. People were getting pretty clever about their hunting techniques, and Paul Martin has made a strong argument for the human hunting hand as more than an incidental factor in the extirpation of many of the mammals of the later Pleistocene.

Martin calls this phenomenon *Blitzkrieg*, or Pleistocene over-kill. Especially when humans are newly arrived on the scene, the sudden appearance of an efficient predator could be a potential disaster for species of game animals. And the plentiful supply of easy pickings might well lead to explosive growth of the hunter population. There are, of course, problems with the straightforward attribution of blame for extinction on humans. (There are always problems with explanations of events of the past, even the relatively recent past.) We know, of course, that large mammals had already disappeared in the Pleistocene—unaided, as far as anyone can tell, by hominid activity. There was a radical change in climate, once again, as that last glacial sheet abruptly retreated (*last* meaning most recent—we are overdue for the next glacial surge, and when and if it comes, it may well overcome whatever contributions we are making to global warming).

And what about all those smaller mammalian species—the voles and mice, for example, many of which also became extinct at about the same time? Humans cannot have, imaginably, hunted them with the same zeal in which we know they pursued bison and may have pursued many of the other grand beasts of the field. Some of the other extinctions, of course, can be tied in with the hunting hypothesis: Humans, certainly, did not systematically hunt dire wolves and saber-tooth cats (though they undoubtedly gladly dispatched them when encountered, if they could). These large carnivores, dependent on the large herbivores, would have followed their prey down the extinction road and, under the *Blitzkrieg* scenario, were simply unwitting, secondary victims of the extirpation of their primary prey. They were, quite plausibly, losers in the new competition with humans for prey. According to Martin, the humans managed to hunt so well that they drove that very set of prey species even unto total annihilation.

Predators generally do not hunt their prey into oblivion. That's not smart; we should recall the classic predator–prey cycle that shows predator population size increasing following a boom in prey. The consequent predator boom then knocks back the prey, but their numbers dwindle so much that the predators suffer a population crash of their own, and so it supposedly goes.

But humans, *hunting* humans, are opportunists, ecological generalists. We will take whatever we can get and our varied techniques are good enough for just about everything. The Bushmen (!Kung) of the Kalahari pursue game of all shapes and sizes, and there is every reason to believe our Upper Pleistocene forebears also went about their hunting business in a most effective manner. So it is not at all beyond the realm of possibility that we will simply hunt any exploitable resource to death. That's what we did with the freshly extinct giant auk (which gamely, if stupidly, walked up the proffered gangplanks in droves, there to serve as provisions while the ship continued its voyage). We also obliterated the late and much lamented passenger pigeons, which as a group were literally shot to death, despite their initial presence in the millions. We know we can (and already have in modern times) hunt species out of existence. Martin simply thinks our progenitors did it in a big way as the Pleistocene was coming to a close.

There is some reason to take his vision seriously. Paleomammalogist Ross MacPhee has documented just how closely late

Pleistocene extinctions correlate with human arrival. Humans got out of Asia and into North America fairly recently. They got to South America fairly soon afterward. But they were somewhat slower in getting to outlying islands such as the Antilles in the Caribbean. And, sure enough, there is a staggered chronology to these late-Pleistocene megafaunal extinctions: The extinctions hit the Old World (even Africa, where the megafauna hangs on, if barely) before they hit the New World. And the large-game extinctions did not seem to occur in any area until humans got there. Another correlation—and another strong inference of cause and effect.

We can never forget climatic input, and the shrinkage and rearrangement of habitat space that really does seem to lie at the very heart of extinctions. Africa lies astride the equator and its megafauna really did survive where others were snuffed out. And humans came from Africa. There was no migration to Africa, and yet the African megafauna was not the first, but rather will be the last, of the Upper Pleistocene megafauna to go.

There is no doubt that as hunters we were involved to some degree in the extinction of some elements of the Upper Pleistocene megafauna. But our input, however severe, strikes me nonetheless as a case-by-case, species-by-species situation. We blatantly and blithely, in modern times, killed off the great auk. But that species, by far the largest and most penguinlike of all auks, could not fly. We did not hunt to extinction the puffins, razorbills, guillemots, and auklets that can fly.

Some humans scattered around the globe, themselves on the brink of extinction, are still hunters and gatherers. Why are they on the brink of extinction? Their habitats are shrinking. Why are their habitats shrinking? Members of technologically advanced societies are changing the nature of their habitats right out from under them. And that is the clue for the real human input into extinction. Hunting has doubtless and very opportunistically gone too far, killing off the prey where, in most carnivore–prey situations, such would never happen. We are just too skilled at taking a wide variety of creatures. And so it has no practical, negative consequences, at least from an energy-procurement point of view, to overhunt, to overkill, to *Blitzkrieg* a few species.

But it was not the technological advancements of modern humans in the hunting arena that got us into the extinction business. Hunters, in most carefully documented hunter-gatherer situations, are not overkillers. Like the Mbuti in Zaire, they are still eminently very much a part of the natural world. It is not hunting, at least as practiced at the subsistence level (the passenger pigeon debacle was mindless slaughter for sport), but rather that consequent technological revolution, the solution to the next energy crisis as Harris puts it, that really got *Homo sapiens* into the extinction racket "big time."

When we settled down and became farmers (superficially less of an arrant bloodthirsty occupation), we actually began to pose an environmental threat that none of our forebears had even remotely come close to: We started to destroy habitat instead of species directly. It is difficult to see our hunting activities as agents of wholesale ecosystem collapse: Hunting contributes more to background than to true mass extinction. But there is no doubt that, with large-scale habitat alteration largely for agricultural purposes, *Homo sapiens* began to mimic the effects that for all of the preceding eons of geological time had been primarily the province of climate change and the odd extraterrestrial bolide.

We have seen that mass extinctions are the direct result of large-scale habitat alteration, and they have happened over and over through purely physical reasons quite without our helping hand. Now, for the first time in the entire 4.55-billion-year history of the earth, *Homo sapiens* are capable of transcending far beyond the incidental, background extinction effects that we, and species such as the brown tree snake in Guam, can and do induce from time to time. For the first time we have a species on earth that has been altering habitat so pervasively that the effects even now border on true mass extinction.

CHAPTER EIGHT

The Canary's Song: Land Use, Habitat Reduction, and Extinction

MADAGASCAR PRESENTS an anomaly in the annals of human history. Our species, moving out of Africa sometime within the last 100,000 years, somehow overlooked this 1,000-mile-long island lying just off Africa's southeastern coast. We got to the far reaches of Asia and Indonesia, following in the footsteps of our ancestor, *Homo erectus* (who made the same trek the better part of a million years earlier), in fairly short order. Humans have been in Australia for at least 40,000 years and in the Americas very likely for at least 20,000, years.

But we went right by Madagascar! Humans were not there until 2,000 years ago. We got there by doubling back: The Malagasy people, an array of some nineteen tribal entities, are nonetheless culturally united in a common language system that shows clear Indonesian affinities. Rice cultivation and elements of religious (notably burial) practices are equally vivid reminders of Malagasy Indonesian origins, though there is a healthy admixture of African-derived customs and physical features of the Malagasy people themselves.

Madagascar's story, then, hinges on a form of double isolation. There was physical isolation throughout the better part of the Upper Mesozoic and Tertiary, allowing a unique range of animals and plants ample time to evolve, and characteristic and unusual ecosystems to become firmly entrenched. Recall that truly large carnivores are absent, as are their usual prey items: The lions and antelope of Africa are simply missing. In their stead is an array of lemurs and ground-dwelling birds that reflect, in part, the relative safety of a terrain free from cats and canids. But the fauna, until 2,000 years ago—well within the time encompassed by written human records—was also isolated from humans. And that lets us evaluate the role humans have played in extinctions in Madagascar's recent past, which in turn should shed some light on the effect that human activity is having on its imperiled biota right now.

Flying over Madagascar's vast and often deeply dissected central plateau, all you see is a monotonous vista of hills and valleys covered with the thinnest of veneers of shrub and grasses. The vegetation is in no way thick enough to mask the deep reddish color of the underlying, hard crusty soil—the clayey laterite so typical of tropical climates. It is poor soil from a farmer's perspective. But such soils are the typical underpinnings of tropical forests the world over. Only the occasional terraced rice fields, and, still rarer, small stands of trees (more than likely imported eucalyptus, rather than native species) interrupt the scene. Both the rice fields and the tree stands tend to be crowded up narrow little valleys on the sides of hills.

It has been supposed for years that the central plateau was originally a sea of tropical vegetation, a forest that joined what is left of the dry forest that still stands in many places in the west with the rainforest that clings to the edges of the steep escarpment that separates the central plateau from the narrow strip of lowland along Madagascar's eastern coast. The supposition has long been that the arrival of humans touched off a wave of cutting and burning that has gone on unabated up to the present moment.

Indeed, the destruction in recent times has been severe: According to a recent report by scientists Glen M. Green and Robert W. Sussman, half of the eastern rainforest has been lost through human deforestation from 1950 to 1985: There are now 3.8 million hectares; there were 7.6 million hectares in 1950, a figure itself reduced from the estimated original (i.e., pre-human occupation) area of some 11.2 million hectares. Green and Sussman point out that destruction has proceeded most rapidly in the flatter and more low-lying areas—places more suitable for agriculture. But recent visits to Madagascar show ample evidence of cutting and burning of even the steepest hillsides on present rainforest margins, and the outlook is totally grim were nothing to be done to intercede with this wholesale destruction of habitat.

Yet, even here in Madagascar, the evidence is not overwhelmingly convincing that deforestation in the past has been solely, or even primarily, the work of humans. Mammalogist Ross McPhee, working with geological, botanical, and anthropological colleagues, has questioned the notion that fire did not come to Madagascar until humans did: There is plenty of evidence of charcoals in deposits going back many thousands of years prior to human arrival. The plants that in many places covered the central plateau seem to have been grasses and the sorts of trees not typically found in true forests. This team has not ruled out good old physical climatic change as a source of vegetational environmental change. There is some evidence that Madagascar's climate has become more seasonal over the last few thousand years—and seasonality, regardless of ambient temperature, calls for more of a eurytopic strategy on the part of resident organisms. Heightened seasonality calls for a strategy more like the eurytopes of the higher latitudes than do the invariant conditions of the tropics. Seasonality could very well have caused the change from forest cover to scrub and grasslands, just as we have seen has been posited for the East African environments, when global temperature drops apparently triggered a radical change from forest to open savannah conditions.

McPhee's interest in Madagascar stems from his pursuit of extinction patterns in island biotas. He reports that, sometime in the last 1,000 years (no one knows exactly when), Madagascar lost a minimum of seventeen of its largest species of mammals, birds, and reptiles. Several species of giant elephant birds roamed the terrain until recently. They were massive creatures that took the theme of flightlessness in Malagasy birds to an extreme that makes ostriches (and even New Zealand's extinct moas) look, if not small, at least emaciated by comparison. There was a pygmy hippo (as I remarked earlier, one of the few mammalian concessions to African propinquity), as well as a species of fouche much larger than that one species of these civetlike carnivores that still prowls Madagascar's forests today. And there were three genera of large-sized lemurs that also just recently became extinct. Largest of all was *Megaladapis edwardsi,* which may have exceeded 120 pounds.

It was invariably the larger animals, especially among the mammals, that so often fell prey to extinction in the Upper Pleistocene. In Madagascar, the latest spate of extinctions took out virtually all large animals. It certainly looks like a case of Paul Martin's *Blitzkrieg:* Humans got there, found out what an easy source of protein these large and, in many cases, slow-moving creatures were, and rapidly hunted them to extinction. But, just as the deforestation scenario on second glance seems not to be solely the act of humans, McPhee and his colleagues are not so sure that the loss of the more than seventeen species of large animals in Madagascar is completely a reflection of overhunting. For one thing, many of the elements of the fauna may have survived nearly to modern times—well beyond the beginnings of regular contact with the Western world.

Early naturalists were greeted with many tales of strange-sounding, large animals, said to be living deep within the country's wild and largely uninhabited (by humans) regions. The animals described sound suspiciously like some of the bigger lemurs, otherwise known strictly from bones unearthed at a few sites. Though these reports themselves could have had an ancient pedigree, it can't be entirely ruled out that these extinct lemurs were indeed still alive until very recently. Madagascar is still poorly explored, at least by eyes trained in Western science. In the 1980s alone, two species of lemur have been discovered and named, and numerous encounters with lemurs and other parts of the biota outside of areas where they have been mapped amply show how incomplete our understanding is of what is really there.

Of course, survival of some of these species until very recently, if that is indeed the case, does not exculpate humans. But it does shift the emphasis from hunting to habitat degradation as the causal connection between human occupation and the disappearance of species. And humans may not have been responsible for the entire transformation of the Malagasy landscape, from essentially forest cover throughout to vast stretches of scrubby arid lands, if indeed forests were ever (i.e., within the last 10,000 years) as extensive as is sometimes assumed. We see what is going on now: massive habitat alteration, as the Malagasy farmers continue to hack away at, and burn down, more and more forest. They assuredly cleared land in the past. They must have hunted at least some of the larger species, as well. And their cats, dogs, and pigs undoubtedly affected the wildlife around them just as they have done everywhere else in the world.

There are good, "natural" ways to explain habitat alteration, and consequent extinction, of Madagascar's larger-body-sized biota in the very recent past. There is also good evidence (including activities that are going on right now) that strongly, if still circumstantially, points the finger at human activity as a strong element that may well have contributed to these extinctions and that is bound to produce many more, as much of the extant fauna seems to be at or very near the brink of extinction. We are clearly at midstream in this latest wave of Malagasy extinctions, and the situation is bound to become worse before it gets better.

This Malagasy microcosm stands for the entire world. Westerners leaving Madagascar more often than not find themselves flying through Charles de Gaulle Airport near Paris, and I need hardly mention that the runways there are surrounded not by mixed deciduous woodlands but by roadways and hectares of cleared farmland. The effect, hectare by hectare, of clearing is greater the closer one is to the tropics, because both microenvironments and entire species are far more narrowly distributed in the tropics than they are in the mid- and increasingly higher latitudes.

Where the best data are available (for serious extinctions in the very recent past, as in Madagascar) there are still uncertainties. But it looks very much, at least to me, that the effects of natural climate shifts on habitats (which, as we have seen, can themselves be clearly responsible for mass extinctions) can as well be exacerbated by human activity. And, attractive as elements of Martin's *Blitzkrieg* hypothesis can be, it is really with the clearing of forests, woodlands, and even desert for use as arable land that humans have begun severely to disrupt the fabric of nature. It is not the effects on this species or that but the degradation of vast tracts of habitat, with varying effects on the species living there, that is by far the most serious source of disruption, the greatest threat of extinction, that humans pose to the rest of the biota.

We need not look very far for the signs, the bellwethers, of extinction. That the canary is quite sickly should by now be apparent to all. Evolutionary biologist E. O. Wilson has estimated that some 17,500 species of terrestrial, tropical organisms face extinction annually. That amounts to between one and two species for every thousand species extant, taking 10 million as a reasonable number of total species on the earth right now. Wilson's estimate is based on the measured rate of destruction of tropical forest habitats taken together with the areas of distribution of tropical species, which, as we saw in chapter 2, are generally rather small.

David Western, of the New York Zoological Society, reports that human activity alone accounts for the destruction of 200 square kilometers of tropical forest and 100,000 square kilometers of rangeland. His estimates of species loss are somewhat more modest than Wilson's: in the hundreds, if not thousands, per year. What is more, Western points out that we are destroying our own habitat: Per year, he reports that 150,000 square kilometers of cropland (i.e., habitat already altered for human economic use) are severely degraded, while 75 billion tons of topsoil are being eroded.

It is difficult to assess whether we are entering a new phase of mass extinction, or merely continuing and exacerbating the Pleistocene story. When the Pleistocene began, ancestral hominid species were subject to the same environmental perturbations as all other species. Our checkered history of extinction and speciation, up at least to the appearance of *Homo erectus,* reads much like the history of other large-bodied terrestrial mammals. Now, when there is no doubt that we ourselves are altering habitats on a massive scale and contributing to the demise of species at what promises to accelerate to a rate far higher than any reasonable rate of background extinction, we must not forget that the climate itself is continuing to change. And separating human from natural climatic effects can be very difficult, to say the least.

Take the issue of global warming. As we saw in chapter 7, atmospheric carbon dioxide has been steadily increasing, matched step-by-step with human population growth. There is no question that the continued accumulation of such gaseous by-products of human activity will, if left unchecked, lead to a net increase in global temperatures through the greenhouse effect: The gases trap solar energy, impeding its reflection off the earth's surface back into space. But how do we interpret the recent spate of mild winters and hotter summers that many North Americans have been experiencing over the last decade? Are we witnessing the first palpable stages of global warming or just normal oscillations occasioned, for example, by nonpermanent shifts in the position of the jet stream? Atmospheric scientists to date cannot decide among themselves, and the disagreement among them is a microcosm of the more general problem of teasing apart the effect of human activity from physical causes of environmental change.

I have already remarked that we seem overdue for another glacial pulse. The famous sixteenth-century Brueghel paintings depicting wintry scenes of skating on frozen canals captured the last miniglacial pulse, the so-called Little Ice Age. Things have been warming up since then, apparently through natural causes. It could be that the latest phase of warming simply continues that trend. Others think that the effects of global warming are being muted, because the natural factors affecting climate would by now be reducing global temperatures, if not for the counterbalancing effect of global warming. Just as we have had a difficult time deciding just how much of the late Pleistocene extinctions really were the direct handiwork of *Homo sapiens* as opposed to climate change, it is not easy in general to tell human from nonhuman agency in the ongoing destruction of the modern biota. Nonetheless the general dimensions and causes of our current predicament seem to me sufficiently clear to warrant the following assessment.

Homo sapiens and the Global Economy of Nature: A Personal Assessment

The most important conclusion of this book is that change of size and location of habitat underlies most extinction events, ranging from isolated species disappearances through the most massive of extinctions. Global climate change, especially global cooling, seems to have been the prime cause of habitat disruption. Now we have added our own activities to the list of physical factors that alter habitat.

It is time to confront some of the more difficult issues that have arisen sporadically in the course of this narrative. How bad is our impact? Do we really need to care about our contribution to species loss? Most importantly, what can we do? After all, we are a sentient species. We ought to be able to analyze our situation and identify steps to remedy the ills revealed by our global miner's canary—to the extent, that is, that the ills are a fallout of human activity. We can never be so vain as to think we are the sole cause of anything in nature, let alone the only agent of extinction.

What follows are my own thoughts on these difficult issues. Scientists are, after all, public citizens, with not only the right but perhaps even the duty to address the political implications of their analyses. To me, these conclusions are straightforward and are reached according to the same canons of reason as, say, my conclusion that extraterrestrial bolides by themselves do not yield a general theory of mass extinction.

Science works by the resolution of conflict. Many issues, including the frequency and importance of bolide impacts as agents of mass destruction, or the relative importance of climatic change to areal habitat change, still have adherents pro and con. I have revealed all sides of these controversies, and I have not shied away from identifying my own conclusions. In that spirit, I give my own thoughts on our current environmental problems, fully aware that my opinions are not the only ones that could be voiced.

Sedentary, agriculture-based human extinction is to the modern world what that asteroid (or swarm of asteroids, or whatever it was that put all that iridium at the K–T boundary 65 million years ago) was to ecosystems at the end of the Cretaceous. We are like loose cannons, able to wreak great damage on our own, and particularly dangerous if our effects happen to coincide with physically induced changes that are also causing extinctions. Global climate change, especially temperature drops or an increase in seasonality, seems more than able to explain the vast majority of extinctions over the past 600 million years. But, as we have seen, there can be a concatenation of factors that may very well render an otherwise difficult situation truly calamitous. Taken in isolation, each factor can have serious effects.

Right now, it is tempting to say that it is primarily global human activity that is driving some species to extinction and threatening many others. As we have just seen, that recent climatic fluctuations of the Upper Pleistocene are still going on and may very well also be contributing to currently observed rates of extinction is harder to see, though I do think that the Madagascar situation points to a concatenation of factors and not just to the human hand. We are too recently arrived to shoulder all the blame for what has been going on for the last 600 million years. But there is equally no doubt that our threat is the greatest one, at least over the short term: We seem to be able to effect more environmental change per unit of time than any other factor ever proposed as a cause for serious bouts of extinction, with the sole exception of the most catastrophic of the bolide impact scenarios.

Extinctions have occurred without any help from us. For the better part of our evolutionary history over the past 4 million years we have been more victims than agents of episodes of extinction. We, as members of our local ecosystems, have been caught up in the events and their consequences of global climate change. But, as the Pleistocene wore on, as our own species emerged, we became more closely involved with these events. Our very culture-based adaptations have taken us beyond the role of mere player in a local ecosystem. We are now a global species. More importantly, we are a global economic system. Our social connections are truly global.

It is hypocritical to blame peoples of emerging, especially tropical, nations for following our own destructive paths of habitat alteration. It is self-delusory not to acknowledge that it is we of the technologically advanced societies who are buying South American beef for our fast-food hamburgers and wood clear-cut from tropical forest stands. Our species, through its economic interconnections, is having a truly global impact. Only the very, very few remaining bands of hunter-gatherers, those not yet tied up with the global economy, retain a life within nature. Ours, increasingly, has been a life without—or against—nature.

The toughest question we can ask at this point is: Is this necessarily a bad state of affairs? Clearly, at some level, I must believe it is, or I would not have bothered to write this book. But it is also true that it is indeed the very nature of our own strategy to deal with the environment, to extract our living from it, to conquer it, to have dominion over it, as the Bible says. We have gotten better and better at it, and there are many who feel that, whatever the consequences to the rest of the physical, let alone the biotic world, we are the better off from it.

As we have seen, George Bush's Interior Secretary Manuel Lujan, making just the latest of a long string of comments of similar ilk from recent U.S. Administration officials, asked if we really need to save every squirrel species, especially if the cost (cost, that is, in terms of dollars of human profit) would be "too great." Taking Mr. Lujan at his word, some spokesmen for development have argued that nature has always replenished species after extinction events, and that we should by no means be concerned by the incidental loss of species in pursuit of the almighty buck.

That, of course, is the long view—that evolution not only replenishes but actually is at its most creative following extinctions. The greater the magnitude of the extinction event, the greater the difference between the organisms that staffed the former ecosystems and those that came along to set up the replacement ecosystems. Extinction resets the evolutionary clock. The issue now is: To what extent is the Western-style, global economy social system that we have fashioned really no longer a part of nature?

The simple truth is that we have indeed altered our place in nature, but we have by no means walked entirely away from it. Local representatives of all other species are integrated into their local ecosystems. Only a handful of human societies still can claim that status. Technologically based societies have overridden this traditional relation between local subdivisions of a species and their immediate environments. But this really doesn't get us completely off the hook: Globally, we are now approaching a concerted whole, a species that by dint of its socioeconomic organization stands as an entity interacting not so much with local but increasingly with the global ecosystem. Rather than each community being a part of a local ecosystem, we interact as single organisms economically primarily among ourselves, and, networked over vast areas, as large-scale social systems with the rest of abiotic and biotic nature.

The emphasis of course is on wresting resources away from the natural world: our goal, according to developers; to our peril, according to conservationists. And it certainly does seem to be very much a one-sided relationship: We take, and thereby despoil and destroy. Harder to appreciate and totally taken for granted are the very real benefits we continue to derive without actively grabbing or destroying anything: the oxygen we breathe, replenished by the photosynthesis of marine plankton and stands of forest; the cycle of water, seemingly so much a given controlled by physical forces, but also very much connected to biological activity within local ecosystems; the stability of the physical environment provided by natural vegetation cover; and the stability of population sizes of various creatures that, unchecked, prove so noticeably annoying. The bug-zappers that glow and snap in suburban backyards can't hold a candle to all those amphibians, reptiles, and birds that devour huge numbers of insects.

For human existence to continue in the technological world, it may seem that the continued existence of any particular species—save a small handful—does not make an iota of difference. But it does make a difference that the ecosystems of the world survive. The global ecosystem is nothing more than the sum of its parts—the local patches of habitat that support local concatenations of energy flow, mixtures of local populations of diverse arrays of plants, animals, fungi, and microorganisms that serve as the local economic machines, continually cycling matter and energy. We have stepped beyond local ecosystems only to find ourselves as a part of the grand global ecosystem. We have not escaped nature—we only think we have, because we have stepped beyond the usual role of integration into local ecosystems.

It is essential for our survival to conserve the global ecosystem, which translates into conserving as much as possible of the natural ecosystems of the world. It isn't really a question of species survival at all (except in our own case). It is quite true that only those of us who love nature will be hurt if the spotted owl of the old forests of the Pacific Northwest is really driven to extinction through destruction of its habitat. Targeting individual species for survival is, in part, an act derived more from aesthetics than economics. But the logging interests are quite right when they accuse conservationists of not wanting to save the owl so much as they want to save the forest itself. The forest—those stands of magnificent Douglas firs and other tree species—stands for the habitat, the ecosystem itself.

Saving endangered species is really the most effective political means yet found to arouse public support for the real task: conserving ecosystems, preserving habitat, putting areas effectively beyond the bounds of exploitation leading to total devastation and destruction. And the argument is purely selfish: to leave enough of the natural ecosystems intact so that the global system remains recognizably what it has been throughout the history of our species so that we as a species can survive. Global changes from natural causes—warming in interglacials, cooling during periods of glacial advance—are in the cards even without our intervention via greenhouse effects derived from our smokestacks and automobile exhausts. With them come growth and shrinkage of ice fields, lowering and raising of sea level, all with effects that, occurring on the time scales of natural climate change, can be coped with a lot more effectively than if we throw our own weight behind the process: If the greenhouse effects really begin to blend with normal global warming, then sea-level rise, for example, can be expected to be many times as rapid as in the past and may prove far more difficult to handle.

We may not need every species of squirrel, but we do need as many—arguably *all*—local ecosystems still remaining to continue to function. Replacing prairie with wheat fields, or diverse tropical forest stands with monotonous eucalyptus growth, just does not do it. The global ecosphere stands only to the extent that its component ecological systems thrive intact. The components of those local systems—local representatives of a vast array of different species—must be kept going. Species act as reservoirs of genetic information, constantly replenishing the organisms that play the critical economic roles in ecosystems that have been downgraded, or even wiped out, by the natural disasters that are always occurring. Species are important as the ultimate source of the players in the ecosystems arena. They keep ecosystems going, but it is the proximate need for us, for our own selfish interests, to keep ecosystems going that really should unite us all in global action to conserve habitat. And that indeed is the longstanding conservationist cry: "Think globally, act locally."

Conservation biologists have expended much time and energy on identifying how much space is required to maintain an ecosystem intact. Often, the issue boils down to how much space is required for some particular species, or small set of species, to be able to survive and continue to reproduce. The problem is tricky, because species vary widely in their areal needs. Not only do species in the tropics differ in general in such requirements from those living in higher latitudes but the variation in needs of the different species staffing local ecosystems within any given region makes it truly difficult to specify any optimum, let alone minimum, size of a reserve—given the assumption that we cannot hope to save an entire habitat area. The solution for each area will depend on a mixture of careful biological analysis, by ecologists and systematists alike, and the political realities of the situation, which is generally the factor that ultimately decides how much habitat really can be set aside. The general rule of thumb, of course, from a biological perspective, always has to be "as much as possible."

That we are still of nature, but on a larger scale removed from the local day-to-day ecosystems concerns that our culture shields us from, seems to be reflected in an emerging sensibility that something is indeed wrong with our economic, adaptive strategy. That sensibility is growing into conscious realization, and, from there, entering the political arena. Greenpeace is the most conspicuously and aggressively active of any number of environmental organizations. Perhaps even more indicative of the growing realization that we are still very much a part of nature is the political side of the green movement: Especially in Europe, green political parties have sprung up and successfully seen candidates elected to national legislative office.

The political problems are enormous. Ecosystem preservation means habitat conservation, which means preventing unbridled human utilization of vast tracts of land. That is okay for those few stretches that have yet to be appreciably populated by humans: Wildlife biologist George Schaller, for example, has been instrumental in establishing what is planned to become the largest wildlife preserve in the world in a remote valley high in the Tibetan Himalayas. But we must be concerned with habitats everywhere. That means telling Malagasy villagers to stop clear-cutting and burning down what little remains of the eastern rainforest. And who are we to tell anyone to do such a thing?

Moralizing and coercion won't do. The time-honored way of resolving economic conflicts is to make the project worth the while of the people whose land is in question. Debt swapping, where a country trades some of its land, agreeing that it be set aside to be conserved, for forgiveness of foreign debts, is one relatively new mechanism that partially achieves this goal. The problem is that the very local people involved usually derive no direct benefit from the swap. They are usually left with that very crucial problem of trying to maintain themselves in face of direct loss of economic resources.

More hopeful policies address this local level directly: Ecologist Dan Janzen, who used his MacArthur Grant to help establish a national park in Costa Rica (a fantastic display of putting one's money where one's mouth is, not to mention one's heart), has involved many local people directly in the workings of the park. Many have become experts on the local fauna and flora, serving as rangers and as guides to the tourists who come to see the birds and other wildlife. They come to see the land as is, without "improvements," as a resource that they can utilize, deriving benefits and all the while protecting it.

If local human populations are not included in the plans, they will always resist, poaching and otherwise illegally disturbing the habitats that they consider their birthright. Nor is tourism always going to supply an effective answer to the problem of involvement of locals in large-scale conservation efforts: There often must be a regulated but non-neglible, continuation of traditional practices. Local farmers the world over are being taught management techniques that allow continued planting around, and sometimes even within, preserved areas—techniques that minimize the negative impacts of farming and allow continued human habitation in the vicinity.

These problems are not exclusively tropical nor restricted to the third world. In the United States, there are always collisions between big business, very much including the local people in their employ, and conservation. The dispute between the logging industry and conservationists rallying around the spotted owl in the forests of the Pacific Northwest is one of the more celebrated cases.

The issue is not strictly one of big business out to exploit natural resources on a grand scale: The Adirondack Mountains of New York State were designated a state park in the nineteenth century. Its 6 million acres make it the largest park in the continental United States. Currently, some 3 million of those acres are owned by the people of the state of New York. The rest are in private hands. Periodically, disputes flare up over who has the say on what can and cannot be done in the Adirondacks. The region, with its poor soil and rough winters, never was much good for farming. Indeed, the wilderness is so impenetrable, so difficult to cope with, that Native Americans apparently never permanently lived there but merely sent in hunting parties from the periphery during the summer. Tanning, logging, trapping and, later, tourism, have been the economic mainstays of the region.

Three groups of people own land in the Adirondacks. There are the big corporations, holding mineral rights, timber reserves, or (and touching off the latest wave of concern) tracts for recreational development (including condominiums). Then there are the second-home people—those who love the mountains, valleys, lakes and streams, the abundant wildlife still to be found, and the deep forests. And then, most important of all, are the permanent residents, most of whom were born and raised there, and all of whom must eke a living from an unyielding terrain and cope with the abundant regulations limiting access to these resources. In general, locals would like more freedom to utilize the land as they see fit (including, but not by any means limited to, sale of lands to developers); outsiders, including many seasonal visitors, would like to see more, not less, regulation and limitation of human use of Adirondack land.

The Adirondacks pose a real and quintessentially human problem—one I broach here to dispel any notion that these issues are germane only to the developing nations of the third world. The Adirondacks represent one of the last great tracts of land to support the summertime breeding of migrant songbirds in the eastern United States. They simply must be kept as pristine as possible. Yet it goes against the grain to arrantly, and arrogantly, overlook the economic rights and needs of the local Adirondack citizenry. It is not only inequitable but it is dangerous: The local people love the land at least as much as those who seek to preserve it from afar. A way simply has to be found to allow limited utilization of the economic resources there to continue and perhaps be expanded a bit in some directions. I have no simple formula for how this might be accomplished, but the example does raise the biggest and most fundamental issue of all.

We have already encountered anthropologist Marvin Harris' notion that human history can be read as a series of successful copings with energy crises. Populations are limited by available resources; each time, in the past, that a given way of life reached its effective population limit, new technologies sprang to the fore, offering what at first invariably seems like unlimited resources. Farming certainly has that effect: Human populations have exploded over the past few thousand years, entirely the result of an ability to feed ourselves more efficiently.

We have this concept of growth, of enhanced exploitation of resources (which invariably results in population growth) that we cannot seem to get away from. Each of us wants more, feels he or she needs more—more money or simply greater opportunity to get more resources. There is no counterbalancing notion of "enough." It seems to me, for example, that in villages peripheral to emerging protected tracts of habitat, a balance should be struck between the current (and traditional) activities of the local peoples and the need to curtail actions that further erode the local ecosystem. The balancing point should involve a concept of enough: Unbridled growth is stopped simply to save the system. But levels of human economic activity sufficient to maintain current levels of human population, but with an explicit realization that those populations should not continue to expand, is probably the most sensible form of compromise that can be reached.

It has been said thousands of times that it is our own unbridled growth—growth in utilization and exploitation of resources, leading to the most important aspects of all, growth of our own populations—that poses the greatest threat to the global ecosystem, and thus, ironically, to our own survival. High population numbers generally help insulate against extinction, but that is for species that have remained integrated into a variety of different local ecosystems.

We have not remained that way; we are global, increasingly interacting with the global ecosystem on a massive scale. Globally, 6, 7, or 10 billion people will have nowhere to hide. It is increasingly unlikely that there will be a vast stretch of the earth housing remnants of our species that can come out from hiding and repopulate the earth should disaster befall us.

Some Final Thoughts

No question the canary is very sick. It looked a bit peaked yesterday, and tomorrow it may well be dead. Migrant song birds, our global canaries, are way down in numbers— alarmingly so in just the past few years. Clear-cutting and burning their winter homes in the tropics is clearly a factor. But so is our continued development here where they come each summer to breed. Not any crummy little wood lot will suffice, with its beer cans and scraggly trees, its neighborhood cats and dogs, nestled in and around suburban developments. Most of these species need larger tracts of relatively unspoiled habitat in which to live and mate. And the birds are our litmus to what is going on in general. It doesn't look good.

We know that extinctions, some extremely severe, have hit in the past. We know that climate change, primarily because it affects size and distribution of habitats, changes the distributions of animals and plants, leading to the extinction of some and the evolution of still others. These things, it is obvious, are natural. They have been going on ever since there has been life on earth. We also suspect that sometimes natural events of the physical environment can be exacerbated by unusual circumstances—as all that iridium at the top of the Cretaceous seems to be telling us. There is no doubt that we, with our ozone depletion, our greenhouse effect, and our clearance of huge areas for agriculture are indeed usurping the role of those Cretaceous asteroids. We are interacting with environmental changes already underway and could well be speeding up the march toward the next mass extinction.

Indeed, we are in the midst of that event right now, judging by the alarming number of species that are disappearing each day. Remember that we measure extinctions by species lost. But the process is actually a matter of ecosystem degradation. I personally, as one of the ever-increasing numbers of fellow lovers of nature, will regret the passing of each and every species. Yet this is not the clear-eyed point, the take-home message. The bottom line here is that the species that we must conserve is our own; we can only do it by realizing that, though the rules have changed, we will never escape intimate relations with the rest of the biosphere. We merely interact more nearly on a global than on a local scale these days. We preserve that global ecosystem by preserving its component parts: not just those in Madagascar and not just those in the Adirondacks but those in New York City as well. Central Park is a fantastic migrant trap and a very important stopover for countless birds each fall and spring. Conserve those habitats, and species—not all, but many—will be saved too.

One of my favorite books is *After Man,* by a whimsical paleontologically trained Scottish painter and writer, Dougal Dixon. He shows us his imagined world 50 million years hence, when *Homo sapiens* has long been extinct and a weird assortment of new animals have evolved and now staff the world's ecosystems. There is much to the claim that it is in the very nature of things for ecosystems to collapse, for species to die off—and for new ecosystems, with new species, to take their place. Never mind that some of those extinction events, such as the Really Big One at the end of the Permian Period, nearly snuffed out life forever. Somehow, life has always managed to survive and reproliferate.

So what's the big deal if it happens again? The great anthropologist Margaret Mead reportedly said to her companion just as she was dying: "Nurse, I think I'm going." The nurse replied, reassuringly: "There, there, dear, we all have to die sometime." Mead is said to have snapped back with her last display of verve and vigor: "Yes! But this is different!" This *is* different. This is not just any ecosystem—it is *our* ecosystem, our own very existence as a species, that is at stake. In the fullness of time, no doubt there will be a replacement biota, whether or not it looks like Dixon's fantasies. But that's later. *This is now.*

Annotated Bibliography
and Suggested Readings

THE LITERATURE on both mass extinctions of the geological past and the mounting threat of extinction facing the modern biota has been growing, seemingly exponentially, over the past decade or so. It is a daunting mountain of information that no one person could possibly surmount.

I have, nonetheless, delved rather deeply into this voluminous pile in the course of preparing this book. Though I have read many of the (generally rather brief) reports that comprise the bulk of the primary scientific literature, I have also made extensive use of several recent compendia that treat aspects of extinction in synoptic, summary form. All such volumes are collections of articles written by scientists who are experts in the particular topic on which their individual contributions are focused.

I shall restrict this list primarily to these sources, indicating in which volume particular authors, mentioned in the text, have contributed such summary articles. I also list additional items (generally books) that bear on extinction or allied topics that I have found particularly useful and illuminating. Together, these sources should provide an entrance to the more technical literature, for all those who wish to delve deeper into the processes underlying the generation and degradation of patterns of biological diversity.

Though this book is written in the spirit that a genuine integration between the phenomena of mass extinctions of the geological past and the events gripping the biota in the here and now is possible, most sources focus on one or the other of these aspects of the extinction problem. For convenience, I follow this separation in the lists below, hoping, however, that the serious student will want to pursue in greater detail topics within both realms of biological inquiry.

Mass Extinction in the Geological Past

Briggs, Derek E. G., and Peter R. Crowther (eds.). 1990.
Palaeobiology. A Synthesis. Blackwell Scientific Publications, Oxford, England. *A single-volume, multi-authored encyclopedia covering a vast range of paleontological topics. M. A. S. McMenamin, P. J. Brenchely, G. R. McGhee, Jr., D. H. Erwin, M. J. Benton, F. Surlyk, L. B. Halstead, and E. L. Lundelius, Jr., provide reviews of specific mass extinction events. Three more general papers address causal issues in mass extinctions: Anthony Hallam examines earthbound causes, especially climate versus habitat area change (he prefers the latter); David Jablonski provides an especially judicious review of the case for extraterrestrial causes; and Jack Sepkoski reviews the case for periodicity of mass extinction, a suggestion that arose from his voluminous computerized data base on the fossil record.*

Donovan, Stephen K. (ed.). 1989. *Mass Extinctions. Processes and Evidence*. Columbia University Press, New York
An excellent summary compendium on mass extinctions of the geological past. Contains an introduction by the editor, three general chapters, followed by nine chapters on particular extinction events. Among these chapters, I have relied particularly heavily on those by S. R. Westrop, P. J. Brenchley, G. R. McGhee, Jr., G. R. Upchurch, Jr., D. R. Prothero, and A. D. Barnovsky, as cited in the text.

Eldredge, N. 1987. *Life Pulse. Episodes from the Story of the Fossil Record*. Facts on File, New York
An overview of the history of life. Examines the effects of extinction on consequent evolutionary history.

Nitecki, Matthew H. (ed.). 1984. *Extinctions*. The University of Chicago Press, Chicago
A collection of eight papers, including "Ecosystem Decay of Amazon Forest Remnants" by Thomas Lovejoy and colleagues. Particularly

*useful to me were the essays by David Raup ("Death of Species"),
Steven Stanley (a critical piece on the role of temperature in marine
mass extinction), and Paul Martin (on his Pleistocene* Blitzkrieg
hypothesis).

Raup, David M. 1986. *The Nemesis Affair*. W. W. Norton, New
York.

*An engrossing and thoroughly entertaining account of the develop-
ment of the theory of extraterrestrial causes of mass extinction, written
by one of the theory's central architects.*

Tattersall, Ian, Eric Delson, and John Van Couvering (eds.). 1988.
Encyclopedia of Human Evolution and Prehistory. Garland Publishing,
New York

*An indispensable source of information on human paleontology, as
well as geology, climates, and paleoecology of the last 5 million years.*

Stanley, Steven M. 1987. *Extinction*. Scientific American Books,
New York.

*A nicely illustrated review of extinctions, and Stanley's views on the
importance of climatic change, especially global cooling, as the causal
trigger underlying mass extinction generally.*

Vrba, Elisabeth S. 1988. Late Pliocene climatic events and hominid
evolution. In Grine, F. E. (ed.), *The Evolutionary History of the
Robust Australopithecines*. Aldine de Gruyter, New York.

*Most of paleobiologist Elisabeth Vrba's work on African antelope
ecology and evolution, as well as her theoretical notions (including the
turnover pulse hypothesis discussed in this book), have appeared only in
technical articles. The paper cited here is an excellent starting place to
learn of her ideas on evolution, diversity, and extinction.*

Extinction in Modern Times

Eldredge, Niles (ed.). 1992. *Systematics, Ecology and the Biodiversity
Crisis*. Columbia University Press, New York.

*Seventeen papers exploring the relation between ecological and system-
atics approaches to understanding biodiversity. Includes a contribution by ecologist George Stevens, of living dead fame. Emphasis is
particularly on the importance of systematics (study of naturally
occurring groups of organisms produced by the evolutionary process)*

233

and the role of museums in the modern biodiversity crisis. Based on a symposium held at the American Museum of Natural History.

Harris, Marvin. 1977. *Cannibals and Kings*. Random House, New York.
Anthropologist Harris sees the major events of human history, especially those leading to significant increases in human population sizes, as the outcome of successful copings with various energy crises. His perspective has much to tell us about the changing position of Homo sapiens *vis à vis both local and, ultimately, global, ecosystems.*

Hoage, R. J. (ed.). 1985. *Animal Extinctions. What Everyone Should Know*. Smithsonian Institution Press, Washington. D.C.
Twelve articles (including one on mass extinctions by paleobiologist Steven M. Stanley) on diverse aspects of the biodiversity crisis. Contains a poignant chapter by anthropologist Colin M. Turnbull on effects of modern civilization on cultural loss preceding actual human extinction—and much food for thought on the relation between hunter-gatherers and their local ecosystems.

Leopold, Aldo. 1949. *A Sand County Almanac*. Oxford University Press, New York.
Eloquent essays by an important mid-twentieth-century champion of conservation—still very much worthwhile reading.

Soulé, Michael E., and Bruce A. Wilcox (eds.). 1980. *Conservation Biology. An Evolutionary-Ecological Perspective*. Sinauer Associates, Sunderland, Mass.

Soulé, Michael E. (ed.). 1986. *Conservation Biology. The Science of Scarcity and Diversity*. Sinauer Associates, Sunderland, Mass.
These two books, with nineteen and twenty-five chapters respectively, provide a comprehensive insight into the connection between biological theory and actual conservation measures. They contain much information on actual diversity patterns, their controls and sources of degradation.

Western, David, and Mary Pearl (eds.). 1989. *Conservation for the Twenty-first Century*. Oxford University Press, New York.
Thirty-three articles, with emphasis on practical conservation measures to combat threats of extinction. Includes several articles explor-

ing the effect of conservation measures on the local economies of indigenous peoples.

Wilson, E. O. (ed.). 1988. *Biodiversity*. National Academy Press, Washington, D.C.
A massive survey of all aspects of the current threat to the world's biota, with heavy emphasis on especially vulnerable regions, and programs and policies devised to thwart ongoing obliteration of habitat and extinction of species.

The Future

Dixon, Dougal. 1983. *After Man*. St. Martin's Press, New York.
Dixon lets his paleontologically informed imagination run wild on what the earth's inhabitants might look like 50 million years after the extinction of Homo sapiens. His paintings of this future bestiary are the highlight of the book, but his message on the effects of extinction on the evolution of life is profound, however unlikely his specific predictions will prove to be.

Index

237

Index

Global warming, 154, 215–216
Gnus, 44
Gorillas, 175
Gould, Stephen Jay, 60, 61, 65, 120, 142
Grand Canyon (Arizona), 3
Grant, Richard E., 50
Graptolites, 81–82
Greenacre, Michael, 44, 46
Green, Glen M., 210–211
Greenpeace, 223
Ground-rollers, 17
Gryphaea, 142
Gubbio (Italy), 105, 107
Guthrie, Dale, 194
Guthrie, Mary Lee, 194

Habitat
  Adirondack Mountains, 224–225
  and Cambrian extinction, 77
  conservation of, 221–224, 228
  and extinction, 10, 111–112, 216
  finding, by species, 9
  in land vs. sea patterns of diversity, 27–29, 40
  loss of, 140
  microhabitats, 40
  size of, and number of species, 166–167
  tracing, 10
  in tropics vs. higher latitude patterns of
    diversity, 27–29, 40
  see also Habitat changes; Madagascar
Habitat changes
  and climatic changes, 194–195
  and environmental changes, 168
  extreme, 133–134
  and *Homo sapiens*, 214–215, 218
  in Madagascar, 213–214
  and mass extinctions, 163–169
Hallam, Anthony, 153–154, 158
Hansen, Thor, 128
Harris, Marvin, 201–202, 207, 226
Hartebeests, 44
Heat energy, 189–190
Herbivores, 14, 101
Higher latitude patterns of diversity.
  See Tropics vs. higher latitude patterns
Holism, 144
"Homage to Santa Rosalia, or Why Are There So
  Many Different Species?" (Hutchinson),
  21
Hominids, 179
  Australopithecus aethiopicus, 184
  Australopithecus afarensis, 180, 181
  Australopithecus africanus, 181–184
  fossils of, 176–178
  history of, 175–176
  Homo habilis, 185
  see also Homo erectus; Homo sapiens
Homo erectus
  bones of, 176–177
  diversification of, 186–187
  in Europe and Asia, 195
  extinction of, 196–197
  and fire, 187
  and *Homo sapiens*, 198
  in Madagascar, 209
  as meat-eaters, 187

and nature, 188
  origin of, 185–186
  success of, 187
  survival of, 196
Homo habilis, 185
Homo neanderthalensis. See Neanderthals
Homo sapiens
  and carbon dioxide increases, 201
  in Caribbean, 206
  and chimpanzees, 179, 199
  and culture, 172–173
  emergence of, 127
  in Europe, 199–200
  and evolution of life, 171
  and extinction, 194, 200–207
  farming by, 202–203
  and global ecosystem, 226
  global impact of, 129, 171–172
  growth of, unbridled, 201, 226
  and habitat changes, 214–215, 218
  and habitat conservation, 224
  and *Homo erectus*, 198
  hunting by, 204–207
  and Late Pleistocene megafauna extinction,
    203
  and nature, 172–174
  and Neanderthals, 59, 172, 197–198
  in North America, 206
  origin of, 199
  and rainforest destruction, 214–215
  and resources, 220
  in South America, 206
  as threat to ecosystems, 18–19
Horn corals, 88
Horses, 29–30
Horseshoe crabs, 31–32
Human brains, 177
Humans. *See* Homo sapiens
Hunting, 42, 204–207
Hutchinson, G. E., 21, 24
Hutton, James, 54

Ice Age
  beetles, 10
  beginning of, 183
  extinction, 16, 42, 129, 131
  gibbons, 176
  Little, 216
  see also Pleistocene Period
Impalas, 44, 45
Invertebrates, 50–51, 108–109
  see also specific types of
Iridium, 97, 105–107, 107, 120, 137, 227
Irish elk, 142–143
Irwin, Douglas, 94
Island biogeography, 166
Isotopes, 53

Jablonski, David (Dave)
  and background extinction, 10, 58, 111, 132
  and Lazarus taxa, 93
  and mass extinctions, 52, 152–153
Janzen, Daniel (Dan), 223–224
Jawless fishes, 68
Johansen, M. B., 114
Johanson, Donald, 178, 180

Index

Index

Index

The Princeton Science Library